A Study in Tinguian Folk-Lore

Fay-Cooper Cole

Alpha Editions

This edition published in 2024

ISBN : 9789364735759

Design and Setting By
Alpha Editions
www.alphaedis.com
Email - info@alphaedis.com

As per information held with us this book is in Public Domain.
This book is a reproduction of an important historical work. Alpha Editions uses the best technology to reproduce historical work in the same manner it was first published to preserve its original nature. Any marks or number seen are left intentionally to preserve its true form.

A STUDY IN TINGUIAN FOLK-LORE

This paper is based on a collection of Philippine folk-tales recently published by the Field Museum of Natural History. [1] The material appearing in that publication was gathered by the writer during a stay of sixteen months with the Tinguian, a powerful pagan tribe inhabiting the mountain districts of Abra, Ilocos Sur, and Norte, of Northern Luzon. In social organization, government, manner of house building, and many other details of material culture this tribe differs radically from the neighboring Igorot. Observation has also led me to the conclusion that the religious organization and ceremonies of this people have reached a higher development than is found among the near-by tribes, and that this complexity decreases as we penetrate toward the interior or to the south. In the main the folk-tales are closely associated with the religious beliefs of the present day, and hence it seems unlikely that they will be found, in anything approaching their present form, far outside the districts dominated by this tribe. Nevertheless, isolated incidents corresponding to those of neighboring peoples, or even of distant lands, occur several times.

In the following pages an attempt has been made to bring together the culture of this people, as it appears in the myths, and to contrast it to present day conditions and beliefs. In this way we may hope to gain a clearer insight into their mental life, and to secure a better idea of the values they attach to certain of their activities than is afforded us by actual observation or by direct inquiry. It is also possible that the tales may give us a glimpse of the early conditions under which this people developed, of their life and culture before the advent of the European.

It should be noted at the outset that no attempt is here made to reconstruct an actual historical period. As will appear later, a part of the material is evidently very old; later introductions—to which approximate dates may be assigned—have assumed places of great importance; while the stories doubtless owe much to the creative imaginations of successive story tellers.

For the purposes of our study, the tales have been roughly divided into three parts. The first, which deals with the mythical period, contains thirty-one tales of similar type in which the characters are for the most part the same, although the last five tales do not properly fit into the cycle, and the concluding story of Indayo is evidently a recent account told in the form of the older relations.

In the second division are the ritualistic and explanatory myths, the object of which seems to be to account for the origin of or way of conducting various ceremonies; for the belief in certain spirits and sacred objects; for the existence of the sun, moon, and other natural phenomena; for the attainment of fire, food plants, birds and domestic animals, as well as of magical jars and beads. Here it should be noted that some of the most common and important beliefs and ceremonies are, so far as is known, unaccompanied by any tales, yet are known to all the population, and are preserved almost without change from generation to generation.

Division three contains the ordinary stories with which parents amuse their children or with which men and women while away the midday hours as they lounge in the field houses, or when they, stop on the trail to rest and smoke.

None of the folk-tales are considered as the property of the tellers, but only those of the third division are well known to the people in general. Those of the first section are seldom heard except during the dry season when the people gather around bonfires in various parts of the village. To these go the men and women, the latter to spin cotton, the former to make fish nets or to repair their tools and weapons. In such a gathering there are generally one or more persons who entertain their fellows with these tales. Such a person is not paid for his services, but the fact that he knows "the stories of the first times" makes him a welcome addition to the company and gives him an enviable position in the estimation of his fellows.

The purely ritualistic tales, called diams, are learned word by word by the mediums, [2] as a part of their training for their positions, and are only recited while an animal is being stroked with oil preparatory to its being sacrificed, or when some other gift is about to be presented to the superior beings. The writer has recorded these diams from various mediums in widely separated towns and has found them quite uniform in text and content. The explanatory tales were likewise secured from the mediums, or from old men and women who "know the customs." The stories of the last division are the most frequently heard and, as already indicated, are told by all. It is evident even to the casual reader that these show much more evidence of outside influence than do the others; some, indeed, appear to have been recently borrowed from the neighboring christianized Ilocano. [3]

TALES OF THE MYTHICAL PERIOD

Reconstruction of the Culture.—In the first division certain actors occur with great frequency, while others always take the leading parts. These latter appear under a variety of names, two or more titles often being used for the same individual in a single tale. To avoid confusion a list of the fourteen principal actors and their relationships are given in the accompanying table. It will appear that there are some conflicts in the use of names, but when it is realized that the first twenty-six myths which make up the cycle proper were secured from six story tellers coming from four different towns, the agreement rather than the disagreement is surprising. As a matter of fact there is quite as much variation between the accounts of the same narrator as between those gathered from different towns.

TABLE OF LEADING CHARACTERS [4]

I. Aponitolau. Son of Pagatipánan [male] and Langa-an [female] [5] of Kadalayapan; is the husband of Aponibolinayen. Appears under the following names: (a) Ligi, (b) Albaga of Dalaga, (c) Dagdagalisit, (d) Ingiwan or Kagkagákag, (e) Ini-init, (f) Ling-giwan, (g) Kadayadawan, (h) Wadagan, (i) Awig (?)

II. Aponigawani. Sister of Aponitolau and wife of Aponibalagen.

III. Aponibolinayen. Daughter of Pagbokásan [6] [male] and Ebang [female] of Kaodanan. Wife of Aponitolau. Appears as (a) Ayo, (b) Dolimáman (?).

IV. Aponibalagen. Brother of Aponibolinayen, and husband of Aponigawani; also appears as Awig.

V. Kanag. Son of Aponitolau and Aponibolinayen. Appears as (a) Kanag kabagbagowan, (b) Balokanag, (c) Dumanau, (d) Ilwisan, (e) also at times is identified with Dumalawi, his brother.

VI. Dapilisan, wife of Kanag.

VII. Dagoláyan. Son of Aponibalagen and Aponigawani. Also appears as Dondonyán of Bagonan—the blood clot child.

VIII. Alokotán. An old woman who acts as a medium. Her home is at Nagbotobotán, where the rivers empty their waters into the hole at the edge of the world.

IX. Gawigawen [male]. A giant who owns the orange trees of Adasin.

X. Giambolan [male]. A ten-headed giant.

XI. Gaygayóma. A star maiden who marries Aponitolau. The daughter of Bagbagak [male], a big star,—and Sinag [female], the moon—.

XII. Tabyayen. Son of Aponitolau and Gaygayóma. Half brother of Kanag.

XIII. Kabkabaga-an. A powerful female spirit who falls in love with Aponitolau.

XIV. Asibowan. The maiden of Gegenáwan, who is related to the spirit Kaboniyan. The mistress of Aponitolau.

In consequence of modern rationalism there is a tendency on the part of a considerable number of the Tinguian to consider these tales purely as stories and the characters as fictitious, but the mass of the people hold them to be true and speak of the actors as "the people who lived in the first times." For the present we shall take their point of view and shall try to reconstruct the life in "the first times" as it appears in the tales.

The principal actors live in Kadalayapan and Kaodanan, [7] towns which our chief story teller—when trying to explain the desire of Kanag to go down and get fruit—assures us were somewhere in the air, above the earth (p. 141). [8] At other times these places are referred to as Sudipan—the term by which spirits are supposed to call the present earth—while the actors are referred to as Ipogau—the spirit name for Tinguian. Whatever its location it was a place much like the present home of this people. The sky, the chief abode of spirits and celestial bodies, was above the land, and the heroes of the tales are pictured as ascending to visit the upper realms. The trees, plants, and animals were for the most part those known to-day. The ocean appears to have been well known, while mention is made of some places in Luzon, such as Dagopan and San Fernando in Pangasinan with which the people of to-day are not at all familiar (p. 89, 168).

We learn that each village is situated near to a river or waterway by the banks of which shallow wells are dug, and there we find the women gathering under the shade of the trees, dipping up water to be carried to their homes, washing and combing their hair, and taking their baths (p. 48). They seldom go singly, for enemies are apt to be near, and unless several are in the company it will be impossible to spread the alarm and secure help in case of attack (p. 43).

Leading up from the spring to the village are bamboo poles on which the heads of enemies are displayed (p. 43). In cases where the warriors have been especially successful these trophies may surround the whole settlement (p. 76). About the town is a defensive wall, generally of bamboo, but in some cases made up entirely of gigantic snakes (p. 43). Within this inclosure are many houses. The bamboo floors are raised high above the ground, while the thatching is of grass. Ladders lead up to little porches, from which doors open into the dwellings. At least part of the houses have a cooking room in addition to that used by the family, while structures containing a ninth room are several times mentioned (pp. 43, 52, 85).

In one corner of the living room is a box containing blankets, above which are pillows and mats used by members of the household and guests; an iron caldron lies on the floor, while numerous Chinese jars stand about. A hearth, made up of a bed of ashes in which stones are sunk, is used for cooking. Above it is a bamboo food hanger, while near by stand jars of water and various cooking pots. Food baskets, coconut shell cups, and dishes, and a quantity of Chinese plates appear when the meal is served, while the use of glass is not unknown. Cups of gold, wonderful jars, and plates appear at times, but seem to be so rare as to excite comment (pp. 33, 98, 102, 105).

Scattered through the village are numerous small buildings known as balaua (p. 43), which are erected for the spirits during the greatest of the ceremonies, and still inside the enclosure are the rice drying plots and granaries, the latter raised high above the ground so as to protect their contents from moisture (pp. 150).

About the town pigs and chickens roam at will, while half-starved hunting dogs prowl about below the kitchens and fight for morsels which drop from above (p. 99). Carabao are kept and used as food (p. 101), but in the cycle proper no mention is made of using them as work animals. [9] Game, especially deer and wild chickens, and fish are added to the domestic supply of food (p. 80), but the staple appears to be mountain rice. Beans, coconuts, oranges, sugar cane, betel-nuts, and tobacco are also cultivated (pp. 33, 107, 121, 138).

Clothing is scanty but nevertheless receives much attention. The poorest of the men wear clouts of banana leaf, and the women, when in danger of capture, don skirts of bark; but on most occasions we find the man wearing a colored cotton clout, above which is a bright belt of the same material, while for ceremonies he may add a short coat or jacket. A headband, sometimes of gold, keeps his long hair in place, and for very special events he may adorn each hair with a golden bead (pp. 74, 76, 81)

The cotton skirts of the women reach from the waist to the knees; the arms are covered with strands above strands of beads, while strings of agate beads surround the neck or help to hold the hair in place. To the real hair is often added a switch which appears to be valued highly (p. 89). Ornaments of gold adorn the ears, and finger rings of the same metal are several times mentioned (pp. 39, 43, 124).

The tales afford us a glimpse of the daily life. In the early morning the chilly mountain air drives the people from their mats to the yard, where they squat about the fires (p. 132). As it becomes light, part of the women begin pounding out the rice from its straw and husks (p. 144), while others depart for the springs to secure water (p. 101). In planting time husband and wife trudge together to the fields, where the man plants the seeds or cuttings, and his wife assists by pouring on water (p. 107). In midday, unless it is the busy season, the village activities are practically suspended, and we see the balaua filled with men, asleep or lounging, while children may be playing about with tops or disk-like lipi seeds (p. 139). As it becomes cooler, the town again takes on life; in the houses the women weave blankets or prepare food, the older women feed the chickens and pigs (p. 93), while the workers from the fields, or hunters with their dogs and game, add to the general din and excitement (p. 80). When night comes on, if it be in the dry season, bonfires spring up in different parts of the village, and about them the girls and women gather to spin. Here also come the men and boys, to lounge and talk (p. 117). A considerable portion of the man's time is taken up in preparation for or actual participation in warfare (p. 74). We have already seen that the constant danger of enemies makes it advisable for the women to go in parties, even to the village spring. One tale informs us of a girl who is left alone to guard the rice field and is promptly killed by the alzado; [10] another states that "all the tattooed Igorot are enemies" (pp. 43, 155, 161).

Revenge for the loss of relations or townspeople is a potent cause of hostile raids; old feuds may be revived by taunts; but the chief incentive appears to be the desire for renown, to be known as "a man who goes to fight in the enemies' towns" (pp. 90, 59).

Warriors sometimes go in parties, sometimes alone, but generally in couples (p. 67). At times they lie in ambush and kill young girls who go for water, or old men and women who pass their hiding place (p. 97). Again they go out boldly, armed with shield, spear, and headaxe; they strike their shields as they go and announce their presence to the enemy (p. 103). In five of the tales the heroes challenge their opponents and then refuse to be the first to use their weapons. It is only when their foes have tried in vain to injure them that they enter the conflict. In such cases whole towns are wiped out of existence and a great number of heads and a quantity of jars and other booty is sent back to the towns of the victors (p. 104). Peace is restored in one instance by the payment of a number of valuable jars (p. 91).

Upon the return of a successful war party, the relatives meet them at the gate of the town and compel them to climb the sangap; [11] then invitations are sent out to friends and relatives in neighboring towns to come and aid in the celebration of the victory (p. 140). When they arrive at the entrance of the village they are met by the townspeople, who offer them liquor and then conduct them to the houses where they feast and dance to the music of gansas (p. 126). [12] Finally the captured heads are stuck on the sagang [13] and are placed by the gate, the spring, and, if sufficient in number, surround the town (p. 140). Taking the heads of one's neighbors does not appear to be common, yet cases are mentioned where visitors are treacherously killed at a dance (pp. 78, 83).

The use of poison [14] is twice mentioned. In one case the victims are killed by drinking liquor furnished by the father of the girl about whose head they are dancing (pp. 148, 156).

Bamboo spears appear to be used, but we are explicitly told that they fought with steel weapons, and there are frequent references to head-axes, spears, and knives (pp. 65, 76, 120).

Marriage appears generally to be negotiated by the mother of the youth at his suggestion (p. 128). At times both his parents go to the girl's home, and after many preliminaries broach the subject of their mission (p. 128). The girl's people discuss the proposition, and if they are favorable they set a day for the pakálon—a celebration at which the price to be paid for the bride is decided upon (p. 49). The parents of the groom then return home after having left some small present, such as a jar or an agate bead, as a sign of engagement (p. 128). [15] The pakálon is held a few days later at the girl's home, and for this event her people prepare a quantity of food (p. 72). On the agreed day the close friends and relatives of both families will assemble. Those who accompany the groom carry jars and pigs, either in part payment for the bride, or to serve as food for the company (pp. 72, 128). The first hours are spent in bargaining over the price the girl should bring, but when this is settled a feast is prepared, and then all indulge in dancing the tadek (p. 59). [16] When the payment is made a portion is distributed among the girl's relatives (pp. 72, 74), but her parents retain the greater part for themselves. [17] The groom cannot yet claim his bride, although in one case he is allowed to take her immediately after the pakálon by making a special payment for the privilege (p. 74). A few nights later the groom goes to the girl's home carrying with him an empty jar with which he makes the final payment (p. 73). The customary rice ceremony [18] follows and he is then entitled to his bride (p. 73). Should the house or anything in it break at this time, it foretells misfortune for the couple, hence precautions are taken lest such a sign should, by accident, be given (p. 60).

In all but two cases mentioned the girl and her husband go to live with his people. In the first instance their failure to do so raises a protest; in the second, the girl's parents are of much more importance than those of the groom, and this may explain their ability to retain their daughter (pp. 138, 159).

When the bride reaches her future home, she sits on the bamboo floor with her legs stretched out in front of her. The slats which she covers are counted and a string of agate beads, equal in length to the combined width of the slats, is given to her. She now becomes a full member of the family and seems to be under the orders of her mother-in-law (p. 60).

The tales give constant sanction for the marriage of near relatives. Dumanau, we are told, marries his cousin, [19] while we frequently meet with such statements as, "We are relatives and it is good for us to be married," or "They saw that they were related and that both possessed magical power, so they were married (p. 35)." It appears that a man may live with his sweetheart and have children by her, yet leave her, and, without reproach, marry another better fitted to be his wife (p. 54). He may also accept payment for a wife who has deserted him, apparently without loss of prestige (p. 64). No objection seems to be raised to a man having two wives so long as one of these is an inhabitant of the upper world (p. 111), but we find Kanag telling his former sweetheart that he cannot marry her since he is now married to another (p. 138). Again, when two women lay claim to Aponitolau, as their husband, they undergo a test and the loser returns to her former home (p. 94). However, this rule does not prevent a man from having several concubines (p., 120). Gawigawen, we are told, is accompanied to a pakálon by eighteen young girls who are his concubines (p. 59).

Divorce is twice mentioned, but it seems to call out protest only from the cast off wife (pp. 63, 149).

Closely associated with the celebration of a marriage seems to be a ceremony known as Sayang, during the progress of which a number of small structures—the largest known as balaua—are built. Judging by their names and descriptions, we are justified in considering them "spirit houses" as they are to-day.

The details of the extended Sayang ceremony are nowhere given, but so much is made plain:—At its beginning many people pound rice, for use in the offerings and for food, and da-eng [20] is danced (p. 40). After the Libon [21] invitations are sent out, by means of betel-nuts covered with gold, to those whose presence is especially desired (p. 62). When the guests arrive at the village spring or gate they are offered food or drink, and then while they dance they are sprinkled with water or rice, after which all go up to the town (p. 41 note 2). A medium who knows the customs and desires of the spirits constructs a bamboo mat, which is known as talapitap, and on it offers food. To call their attention she frequently strikes the ground with the dakidak—split sticks of bamboo and lono [22] (p. 40). The guests are not neglected, so far as regards food, for feasting and dancing occupy a considerable portion of their time. The ceremonial dance da-eng is mentioned, but the tadek [23] seems to be the one in special favor (pp. 41, 59).

One tale tells us that the Sayang was held immediately following a head hunt; and another, that Aponitolau went out to get the head of an old man before he started this ceremony (pp. 69, 76); however, the evidence is by no means conclusive that it is related to warfare.

On page 105 we are told that Kanag's half sister is a medium, and the description of her method of summoning the spirits tallies with that of to-day. At the Sayang ceremony she is called to perform the Dawak, [24] with the assistance of the old woman Alokotán (p. 106). The Dawak is also held in order to stop the flow of blood from Aponitolau's finger (p. 113). The only other ceremony mentioned is that made in order to find a lost switch (p. 91).

Certain well-known customs are strongly brought out in our material. The first, and apparently most important, is the necessity of offering liquor and food, both to strangers and to guests (p. 58). Refusal is so keenly resented that in one instance a couple decline to allow their daughter to marry a man whose emissaries reject this gift (p. 73). Old quarrels are closed by the tender of food or drink, and friendships are cemented by the drinking of basi [25] (p. 134). People meeting for the first time, and even friends who have been separated for a while, chew betel-nut together and tell their names and places of residence. We are repeatedly told that it is necessary to chew the nut and make known their names, for "we cannot tell our names unless we chew," and "it is bad for us if we do not know each other's names when we talk." A certain etiquette is followed at this time: old men precede the younger; people of the home town, the visitors; and men always are before the women (pp. 45, 133). The conduct of Awig when he serves liquor to the alzados [26] is that of to-day, i.e., the person who serves always drinks before passing it to others (p. 156).

Certain other rules of etiquette or restrictions on conduct come out in the tales. We learn that it is not considered proper for a man to eat with the wife of another during his absence, nor should they start the meal before he comes in (p. 52). The master of a dance is deeply chagrined and chides his wife severely, because she insists on dancing before he has invited all the others to take their turns (p. 70). Greediness is reproved in children and Aponitolau causes the death of his concubines whose false tales had led him to maltreat his wife (p. 116). Unfaithfulness seems to be sufficient justification for a man to abandon his wife and kill her admirer (p. 78); but Kanag appears as a hero when he refuses to attack his father who has sought his life (p. 121).

Of the ceremonies connected with death we learn very little except that the women discard their arm beads, the mourners don old clothing, and all wail for the dead (pp. 44, 90). Three times we are told that the deceased is placed on a tabalang, or raft, on which a live rooster is fastened before it is set adrift on the river. In the tales the raft and fowl are of gold, but this is surprising even to the old woman Alokotán, past whose home in Nagbotobotán all these rafts must go (p. 131).

Up to this time in our reconstruction of the life of "the first times" we have mentioned nothing impossible or improbable to the present day Tinguian, although, as we shall see later, there are some striking differences in customs and ideas. We have purposely left the description of the people and their practice of magic to the last, although their magical practices invade every activity of their lives, for it is here that the greatest variations from present conditions apparently occur.

These people had intimate relations with some of the lesser spirits, especially with the liblibayan, [27] who appear to be little more than their servants, with the evil spirits known as banbanáyo, [28] and with the alan [29] (p. 123). The alan, just mentioned, are to-day considered as deformed spirits who live in the forests: "They are as large as people but have wings and can fly; their toes are at the back of their feet and their fingers point backwards from their wrists." The several references to them in the tales such as "you alan girls whose toes on your feet turn out" indicate they were so considered in the first times (p. 161). Some of them are addressed as "you alan of the springs," and in one instance a man dives down into the water where the alan live (p. 148), but in general their homes seem to be similar to but much finer than those of the people of Kadalayapan and Kaodanan. These spirits appear time after time as the foster mothers of the leading characters: Generally they secure a drop of menstrual blood, a miscarriage, or the afterbirth, and all unknown to the real parents, change them into children and raise them (p. 83). These foster children are pictured as living in houses of gold situated near springs, the pebbles of which are of Gold or beads; [30] the places where the women set the pots while dipping water are big plates or dishes, while similar dishes form the stepping stones leading up to the house. Articles of gold are found in the dwellings and valuable jars are numerous. When the true relationships of these children are established they always go to their blood parents, carrying with them these riches, which are a source of wonder and comment (pp. 43, 64).

The people of Kadalayapan and Kaodanan have many dealings with the celestial bodies. The big star Bagbagak appears as the husband of Sinag—the moon—and father of the star maiden Gaygayóma, who, Aponitolau assures his wife, is a spirit. When this girl comes down to steal sugar-cane she takes off her star dress and appears as a beautiful maiden; [31] she becomes enamored with Aponitolau and takes him to the sky, where he lives with her. They have a child, who later marries in Kadalayapan and thereafter stays below. Upon the occasion when Aponitolau visits his first wife and fails to return to the sky at the appointed time, a great company of stars are sent to fetch him, with orders to devour him if he refuses to obey (p. 109, ff.).

In the first tale Aponitolau himself appears as "the sun," "the man who makes the sun," as "a round stone which rolls," but when it is established that he is the son of a couple in Kadalayapan he apparently relinquishes his duties in the sky and goes to live in the village of his people. With him goes his wife Aponibolinayen, who had been carried above by a vine. While at his post in the heavens, Aponitolau is closely associated with the big star, whose duty it is to follow him in the sky. Again we are told that Aponitolau is taken up by the spirit Kabkabaga-an, whom he marries and by whom he has a son (p. 114). In some instances this hero and his son Kanag converse with thunder and lightning, which appear at times not unlike human beings (p. 100); but in the eighth relation the two kinds of lightning are pictured as dogs who guard the town of Dona.

These people enjoy unusual relations with inanimate things, and we find them conversing with spears and with jars. [32] In one case the latter appear to be pastured like animals, and surround Aponitolau when he goes to feed them with lawed [33] leaves and salt (p. 51). Weapons weep blood and oil when taken down for the purpose of injuring certain persons (p. 43). A nose flute, when played by a youth, tells him of his mother's plight (p. 152), while a bamboo Jew's harp summons the brothers of its owner (p. 162). Animals and birds are frequently in communication with them: The hawk flies away and spreads the news of the fight at Adasin [34] (p. 90); at the bidding of Dalonágan a spider spins a web about the town (p. 124); and Aponitolau is enabled to fulfill the labors assigned him by the ten-headed giant only through the aid of spiders, ants, and flies (p. 101). [35] During certain dances the water from the river flows over the town and fish come up and bite the feet of the dancers (p. 59). Crocodiles are left to guard the sister of Aponibalagen, and when they fail to explain their negligence they are whipped and sent away by their master (p. 87). A great bird is pleased with Aponitolau and carries him away [36] to its home, where it forces him to marry a woman it had previously captured (p. 92). In one instance an animal gives birth to a human child; a frog laps up the spittle of Aponitolau, and as a result becomes pregnant [37] and gives birth to a maiden who is taken away by the spirits (p. 105). Another account states that the three sons of Aponitolau and Aponibolinayen are born as pigs, but later assume human form (p. 116). Kanag becomes a snake when he tries to secure the perfume of Baliwán, but is restored to human form when he bathes in a magic well (p. 137). These and other mysterious happenings, many of which are not explained as being due to their own volition, befall them; thus Ingiwan, while walking, is confronted by an impassable hill and is compelled to cross the ocean, where he finds his future wife, but upon his return the hill has vanished (p. 86). In other instances the finger rings of people meeting for the first time exchange themselves (p. 92). The headband of Ligi flies away without his knowledge and alights on the skirt of a girl who is bathing in the river. As a result she becomes pregnant, and when the facts become known Ligi is recognized as the child's father (p. 144). It seems probable that the superior powers are responsible for these occurrences, for in at least one instance the great spirit Kaboniyan steals a maiden and turns her into a flock of birds, who talk with and assist the owner of a rice field (p. 151).

While they thus appear to be to a certain extent under the control of the spirits and to be surrounded by animals and inanimate things with human intelligence and speech, the people of these "first times" possess great power over nature: Time and space are annihilated, for at their will daylight comes at once (p. 150), or they are transported to a place in an instant (p. 92). At their command people appear: Kanag creates betel-nut trees, then cuts the fruit into bits, Which he sows on the ground. From these come many people who are his neighbors, and one of whom he marries (p. 121). The course of nature is changed: A field is planted in an instant; the crops mature in a few days, and the grain and fruits take themselves to the storehouse (p. 150). A strike-a-light turns into a hill which impedes pursuers (p. 75), [38] while a belt or head-axe serves as a ferry across a body of water (p. 84). A storm is called upon to carry a person or a building to a distance (p. 121), and a spring is created by killing an old man (p. 60). [39] Prepared food appears at a word; a stick when cooked becomes a fish, and though it is repeatedly broken and served it always appears ready for service at meal time (p. 33); a small jar containing a single grain of rice supplies an abundance of food; another jar no larger than a fist furnishes drink for a company and still remains a third full; while a single earring fills a pot with gold [40] (pp. 47, 119, 123).

Quite as easy as the creation of beings is the causing of sleep or death. All the people of a village are put to sleep at the will of a single person (p. 145) and Albaga—while still at a distance—causes the death of Aponibolinayen (p. 44). At a word of command the spears and head-axes of the people of Kadalayapan and Kaodanan go out and kill great numbers of the enemy, and the heads and booty take themselves in orderly fashion to towns of their new owners (pp. 66, 75). Many methods of restoring the dead to life are employed; spittle is applied to the wounds, or the victim is placed in a magic well, but the common method is for the hero "to whip his perfume," [41] whereupon the dead follow his commands (pp. 152, 157).

The birth of a child, to a woman of these times, is generally preceded by an intense itching between the third and last fingers, and when this spot is pricked the child pops out "like popped rice." [42] Its growth is always magical, for at each bath its stature increases by a span (p. 102). Within a few days the baby is a large child and then begins deeds of valor worthy of the most renowned warriors (pp. 95, 96).

The power of assuming animal forms appears to be a common possession, and we find the different characters changing themselves into fireflies, ants, centipedes, omen birds, and in one case into oil [43] (pp. 85, 99).

One of the most peculiar yet constantly used powers of these people is their ability to send betel-nuts on various missions. Whenever an invitation to a ceremony or celebration is to be extended, nuts covered with gold are oiled and sent out. They go to the intended guest, state their errand, and, if refused, forthwith proceed to grow on his knee, forehead, or pet pig, until pain or pity compels him to accept (p. 146). In some cases it appears that the nuts themselves possess the magic properties, for we find Aponitolau demanding that his conquered foes give him their betel-nuts with magic power (p. 91).

Relationships can be readily ascertained by the chewing of these nuts, for when the quids are laid down they are transformed into agate and golden beads and lie in such a manner that the associations are fully established (pp. 35, 36, 41).

Enough has been mentioned to show how important a part magic and magical practices play in the life of this people, but one further reference should be made, since it is found in nearly every tale. When the marriage price is settled upon, the mother of the groom exercises her power and at once fills the spirit house with valuable jars and the like; this is repeated until enough are gathered to meet the demands of the girl's people (p. 133). Even when the agreed sum has been delivered we often find the girl's mother herself practicing magic, to secure additional payment, and by raising her elbows or eyebrows causing a part of the jars to vanish (pp. 133, 143).

Despite their great gifts we find that these people are not all-powerful and that they deem it wise to consult the omens before starting on a task or a journey. The gall sack and liver of a pig are eagerly examined, [44] while the calls of birds, actions of animals, or signs received from the thunder and lightning regulate their conduct. In cases where these warnings are disregarded misfortune or death always overtakes the individual (pp. 48, 49, 100 ff).

Death comes to them, but apparently is only a temporary state. The deceased are often revived by some magical process (p. 152), but if not the corpse is placed on a raft and is set adrift on the river. [45] The streams and rivers, we are told, all flow past Nagbotobotán before they empty into the hole where all streams go. In this place lives the old woman Alokotán, who is related to the people of Kadalayapan and Kaodanan. Her duty it is to watch for dead relatives, to secure them, and make them alive again (p. 132). She is the owner of a magic pool, the waters of which revive the dead and renew youth.

Comparison of the Reconstructed Culture with Present Day Conditions.—Before passing to a consideration of the tales in the last two divisions of our material, it may be well to compare the life and beliefs of these "people of the first times" with those of the living Tinguian. Kadalayapan and Kaodanan appear, in a vague way, to have been located in Abra, for we learn that the Ilocano, Don Carlos, went up the river from Baygan (Vigan) [46] to Kadalayapan; that the alzados [47] lived near by; while the tattooed Igorot occupied the land to the south (pp. 77, 155). The villages were surrounded by defensive walls such as were to be found about all Tinguian villages until recent times, and which are still to be seen about Abang and other settlements. Within the walls were many houses, the descriptions of most of which would fit the dwellings of to day. The one thing which seems foreign to present conditions is the so-called "ninth room" which receives rather frequent mention. There is nothing in the tales referring to buildings or house construction which lends support to the contention of those who seek to class the Tinguian as a modified sub-group of Igorot. [48] The Bontoc type of dwelling with its ground floor sleeping box and its elevated one room kitchen and storage room is nowhere mentioned, neither is there any indication that in past or present times the Tinguian had separate sleeping houses for the unmarried men and boys, and for the girls, as do their neighbors to the south.

The other structures, such as the spirit houses, rice drying frames, and granaries were similar to those seen to-day in all the villages. Likewise the house furnishings, the musical instruments, and even the games of the children were such as are to be found at present, while our picture of the village life given on page 6 still fits nearly any Tinguian settlement in Abra. The animals mentioned are all familiar to the present people, but it is worthy of note that in the first twenty-six tales, which make up the cycle proper, the horse is not mentioned, nor does the carabao appear to be used as a work animal. Still more important is the fact that the terraced fields and the rice culture accompanying them, which to-day occupy a predominant place in the economic life of the people, are nowhere mentioned. On the other hand, the langpádan, or mountain rice, assumes a place of great importance. References to the cultivation of the land all seem to indicate that the "hoe culture," which is still practiced to a limited extent, took the place of agriculture.

The clothing, hair dressing, and ornaments, worn by these people, agree closely with those of to-day. Beads seems to have been of prime importance, but could scarcely have been more prized or more used than at present. Unless she be in mourning, the hair and neck of each woman are now ornamented with strings of beads, many of them of evident antiquity, while strands above strands cover the arms from the wrist to the elbow or even reach to the shoulder. [49]

The wealth of a person seems to have been, to a large extent, determined by the number of old jars in his possession. As at the present time, they formed the basis of settlement for feuds, as payment for a bride, and even figured in the marriage ceremony itself. The jars, as judged from their names, were evidently of ancient Chinese manufacture, and possessed power of speech and motion similar to that of human beings; but in a lesser measure the same type of jars have similar powers to-day. [50]

The use of gold and jewels seems to have been common in the old times; the latter are seldom seen in the district to-day, but the use of bits of gold in the various ceremonies is still common, while earrings of gold or copper are among the most prized possessions of the women. [51] Placer mining is well known to the Igorot of the south, who melt and cast the metal into various ornaments. So far as I am aware, this is not practiced by the present Tinguian, but may point back to a time when the industry was known in this region, or when trade relations with the south were much freer than in recent years.

The weapons of the warriors, which we are specifically told were of metal, are identical with those seen at the present time, while the methods of warfare agree with the accounts still told by the old men of their youthful exploits.

A survey of the tales brings out boldly the fact that a headhunt was one of the most important events in Tinguian life. To-day stress of circumstances has caused the custom to suffer a rapid decline, but even now heads are occasionally taken, while most of the old men have vivid recollections of the days when they fought "in the towns of their enemies." A spirited account of a head celebration seen in the village of Lagangilang—from which ten of these tales were collected—will be found in the writings of La Gironiere, already referred to. [52] It is important to note that this account, as well as those secured from many warriors of the present generation, offers some striking differences to the procedure in the olden days, particularly as regards the disposal of the skulls. The tales tell of the heads being placed on the sagang [53] at the spring, at the gate, or about the town, after the celebration. Certain of the present villages make use of the sagang, but the more common type of head holder is the saloko, [54] which still figures in many ceremonies. However, the heads only remain in these receptacles until the day set for the festival. They are then carried to the centre of the village and there, amid great rejoicing, are cut open; the brains are removed and to them are added the lobes of the ears and joints of the little fingers, and the whole is then placed in the liquor, which is served to the dancers. Before the guests depart the skulls are broken into small pieces and a fragment is presented to each male guest, who carries it home and is thus often reminded of the valor of the takers. [55] A study of Tinguian beliefs furnishes an additional religious motive for the taking of heads, but with the people of Kadalayapan and Kaodanan revenge and the desire for renown were the prime incentives.

Every tale emphasizes the importance of the Sayang ceremony and the spirit structure known as balaua. [56] The ceremony is nowhere described in full, but the many details which are supplied show that it was almost identical with that of to-day. The same is true of the Dawak, [57] which we find mentioned on three different occasions, and of the ceremony made to aid in locating lost or stolen articles. The most noticeable fact, to the person familiar with Tinguian life, is that these are the only ceremonies mentioned among the many known and practiced at present. More than a score of different rites are now well known to this people, and occupy a very considerable portion of their time and attention during the first four months of the year.

The failure to make mention of these very important events is explained, it seems to me, not by their absence, but by the fact that these rites vary in importance and that the privilege of celebrating them is hereditary in a family. Should one not entitled to hold such a ceremony desire to do so, he must first give, in order, all the lesser events, a costly procedure extending over a period of several years. The people of Kadalayapan and Kaodanan always appear as being closely related to the spirit Kaboniyan, [58] and exceedingly powerful. It seems probable that the story teller takes it for granted that all of them are entitled to hold the most important ceremony known to the Tinguian.

A prominent figure in these rites is the medium, through whom the ancient people generally conversed with the spirits, but in exceptional cases we found the heroes talking direct with the superior beings; however, this gift is not confined to the men of old, for in such tales as 55 and 59 people who are believed to have lived recently have conversed with the spirits and have even been joined to them in marriage.

The procedure in choosing a bride, the engagement, the pakálon, [59] and the marriage proper are all those of the present day, but the rules governing the marriage of relatives differ radically. As already noted, one of the chief qualifications for marriage, among the people of the tales, was relationship, and even cousins became husband and wife. Such a thing is unthinkable among the Tinguian of to-day; first cousins are absolutely barred from marrying, while even the union of second cousins would cause a scandal, and it is very doubtful if such a wife would be allowed to share in her deceased husband's property. [60]

It appears that only one real [61] wife is recognized as legitimate, but that from "the first times" to the present a man might have as many concubines as he could secure.

So far as mythology and present day conditions can inform us the bride has always gone to the home of her husband and, for a time at least, has been subject to the dictations of her mother-in-law, although the couple are generally soon established in a home of their own, in the town of the groom. There is nothing in Tinguian life or tradition to indicate that they have ever had a clan system or a matriarchal form of government.

The few references to the procedure immediately after a death indicate that, in part, the people of to-day follow the old custom; but here again an important departure occurs. We are thrice told that the corpse was placed on a little raft called tabalang and set adrift on the river; and in one case the afterbirth was treated in the same manner. Nothing of the sort is done to-day, nor does it seem at all likely that such has been the case in recent generations. The body is now buried beneath the house, and certain set rules govern the movements of all persons related to the deceased, as well as the disposal of the corpse. This procedure is so complex and so uniform throughout the whole Tinguian belt that it seems improbable that it has grown up, except through a long period of time. At this point it is interesting to note that at many ceremonies it is necessary to construct a small raft called tal-talababong, or talabong, to place offerings in it, and set it adrift on the stream, in order that any spirits who have been prevented from attending the ceremony may still secure their share. [62]

The festivals, the dances, the observances of the proprieties required by good breeding or custom of to-day, follow closely those given in the tales. The greatest divergence is in the offering of betel-nuts and the telling of names, which occupies such an important place in the narratives. The use of betel-nut for chewing is less common among the Tinguian people than with most other Philippine tribes, a fact which may be accounted for by their constant use of tobacco. However, betel-nuts still occupy a most important place in the various ceremonies, and many offerings intended for the spirits must be accompanied with the prepared nut. In nearly every instance when invitations were sent out, for a ceremony, the people of the tales intrusted an oiled betel-nut covered with gold with this duty. This has its counterpart to-day in the small gifts of gold which are often carried to some friend, in another town, whose presence is particularly desired. It seems not improbable that the golden colored husks of the ripe betel-nuts may have suggested the substitution.

Magic was practiced extensively in "the first time," but it is by no means unknown to the people of the present day. They cannot now bring a dead person to life, or create human beings out of bits of betel-nut; but they can and do cause sickness and death to their foes by performing certain rites or directing actions against garments or other objects recently in their possession. Even the name of an enemy can be applied to an animal or inanimate object and action against it be transferred to the owner.

Like the Tinguian, the people of Kadalayapan and Kaodanan are warned or encouraged by omens received through the medium of birds, thunder, lightning, or the condition of the gall and liver of a slaughtered pig; [63] and like them they suffer for failure to heed these warnings, or for the infraction of a taboo.

The myths of the first division make it plain that, to the people of those times, the sun, moon, and stars were animate—either spirits or human beings. In some cases a similar conception was held for thunder and lightning, while in others they appear as animals. It will appear that such ideas are not foreign to the second division of the tales, which represent present day beliefs. Thus, in the mountain village of Baay the sky is considered as a male spirit—the husband of the earth, and father of sun and moon. Again, in Lagangilang and Abang, the thunderbolt is identified as Kadaklan—the most powerful of all spirits—who "often eats the ground and releases his wife Agemem."

This brings us to a most interesting question, namely: Are the chief actors in our tales to be considered as celestial beings and spirits, or as human heroes? We have already made note of the fact that in the first tale Aponitolau is identified with Ini-init whom, we are told, was "the sun," "the man who makes the sun," "a round stone which rolls." In this tale he marries Aponibolinayen, a maiden whose name may possibly be construed to mean "the woman in the moon." [64] However, we find Aponitolau abandoning his place in the sky and going to reside in Kadalayapan. This tale comes from the town of Langangilang where, as we have already seen, the celestial beings are regarded as spirits. Tale fifteen, coming from the same town, shows us this same Aponitolau going up to the sky, where he marries the spirit Kabkabaga-an, but as before he returns to his home below. A further indication of his celestial character is perhaps afforded us in tale fourteen, which was recorded in Patok, a valley town in which the sun, moon, and stars are now regarded as "lights" belonging to the spirit

Kadaklan. Here we find that Aponitolau marries the star maid Gaygayóma, who is the daughter of the big star Bagbagak, and Sinag—the moon. In this same tale Aponibolinayen appears as the first wife of Aponitolau, and it is clear that in the mind of the story teller she is not identified with Sinag. Aponitolau appears in the other tales without any hint of celestial qualities. Aside from her name and the fact that she is once pictured as visiting the sky, there is nothing to indicate that his wife Aponibolinayen is to be considered as the moon. A careful study of the other characters who reside in Kadalayapan and Kaodanan fails to yield any evidence that they are considered as celestial beings.

During the Sayang ceremony held in San Juan, a certain man and woman, who are then called Iwaginán and Gimbagon, [65] represent the good spirits and are defended by the people when evil spirits try to dispossess them of their property. This is the only instance I have observed in which the names of any of these characters of the tales appear in the ceremonies, while a list of more than one hundred and fifty spirits known to the Tinguian fails to reveal more.

While in the practice of magic, and in their communication with nature, celestial bodies, and spirits, these "people of the first times" far excelled the present Tinguian, they had a material culture and ceremonial life much like that still found in Abra.

It seems then that these people, about whom the stories cluster, are not to be identified as celestial beings or spirits. [66] They appear rather as generalized heroes whose life and deeds represent that of an earlier period, magnified and extolled by succeeding generations.

RITUALISTIC AND EXPLANATORY MYTHS

The second division of the tales now assumes a position of importance to us, for in it we find present day ideas and beliefs of the people strongly brought out, and are thus in a position to contrast them with the tenets of the people in "the first times."

The influence of custom is exceedingly strong among the Tinguian of today. The fact that the ancestors did so and so is sufficient justification for performing any act for which they have no definite explanation. Nowhere is this influence greater than in the ceremonies. These, which accompany all the important happenings in their daily life, are conducted by mediums who are fitted for office by long training, and each one of whom is a check on the others if they wilfully or through carelessness deviate from the old forms. The ritual of these ceremonies is very complex and the reason for doing many acts now seems to be entirely lost, yet the one explanation "kadaúyan"—custom—is sufficient to satisfy any Tinguian. Other acts, as well as the possession of certain things, are explained by myths, such as we are considering. It seems certain that we are here dealing not with present day beliefs alone, but with at least relatively old customs and tales, which while enabling us to understand present day conceptions also give us a glimpse into the past.

The myths 32-40, which are known to the people as diams, are now inseparable parts of the various ceremonies. Thus, when a pig is to be offered in the Sayang ceremony, the medium sits down beside it and strokes it with oiled fingers while she "talks to the spirits." The translation of her "talk" shows that this is in no sense a prayer but is rather an account of how the greatest of the spirits taught the Tinguian people to perform this ceremony correctly. Likewise, when she offers food in the Dawak [67] ceremony, she relates how the spirit Kaboniyan taught the Tinguian to do this in the same manner that he performs it. In the Pala-an [68] diam she relates, in story form, the cause of the sickness, but in this case ends with a direct invocation to the spirits in Dadáya to "make them well again if you please." The balance of the diams, 35-40, are in story form, and seem intended more as an explanation to the people as to the causes of their troubles than to be directed toward the spirits. However, the medium seldom has an audience, and rarely ever a single listener, as she recites the diams she has learned verbatim from her instructors when preparing for the duties of her office.

Myths 41-54 are of quite a different type. They are generally told by the mediums or wise old people, during the ceremonies, but always to a crowd of eager listeners. They are not learned word for word, as are the diams, but their content is constant and they are thoroughly believed.

That they exert a great influence on the beliefs and conduct of both old and young is undoubted. The evil which befalls a person who molests the guardian stones is thus made known even to the children who generally keep at a distance from the grove in which they stand. Again, these tales give sharp warning as to what befalls a person who even ignorantly breaks the taboos following a death; but at the same time advance means of thwarting the wrath of the enraged or evil spirits.

Myths 55 to 62 at first glance to not appear to be explanatory at all, but seem rather to be a series of stories dealing with the relations between certain persons and the natural spirits or those of the dead. However, it is the intent and use rather than the form of these stories which has caused them to be included in this division, for they give the people authority for certain beliefs and conceptions which they hold. Tale 56 gives us a glimpse of the prevalent idea of the abode of the dead, where the spirits lead much the same sort of life as they did while alive, but we secure quite a different picture of this realm from the Baluga [69] tale, in which the home of the deceased is said to be in the ground while the "life" of the dead woman is kept in a bamboo cup. This last account was heard in Manabo, a town near to the Igorot settlements of the Upit river, and may be influenced by the beliefs held in that section. [70]

Certain individuals appear to have intimate dealings with the natural spirits, in some instances even being joined to them in marriage. The afterbirth child, Sayen, is believed to have lived "not very long ago;" yet we find his life and actions quite similar to those of the heroes in "the first times," while his foster mother—the alan [71]—takes the same part as did the alan of old.

Relations 63 to 74 appear as pure explanatory tales, accounting for the existence and appearance of celestial bodies and animals in their present state; they also account for the possession of fire and of many prized objects, such as jars and agate beads. Incidentally many essential traits and old customs come out, such, for instance, as those of war and mourning, which appear in connection with the origin of the kalau. [72]

With few exceptions the myths of this division correspond to present beliefs; the spirits are those known to-day; the towns mentioned are now existing or their former locations are well known. They have thus the appearance of being of more recent origin than those of the first division, yet it is worthy of note that there is little in them which seems foreign to or out of keeping with the older tales.

FABLES

The last division may be said to be made up of fables, for the story tellers without hesitation label them as fictions. The last of these appears to be only a worked over incident of myth 56, in which the big bird Banog carries the hero to its nest, from which he escapes by holding to the wings of the young birds. It is possible that more of these fables are likewise incidents in tales prevalent among the Tinguian, but not heard by the writer. Whether or no this be true, it is certain that most of these stories are well known to the Ilocano of the coast and the other Christianized natives throughout the archipelago. Comparison with the folk-lore from other regions shows that these stories are by no means confined to the Philippines. The chief incidents in the narrative of the turtle and the monkey have been recorded from the Kenyah of Borneo [73] and from the northern peninsula of Celebes; [74] the race between the shell and the carabao is told in British North Borneo [75] in regard to the plandok and crab, while it is known to European children as the race between the turtle and the hare. The threat of the mosquito in 84 is almost identical with that recorded by Evans in Borneo; [76] while many incidents in the fable of Dogidog [77] are found in the Iban story of Simpang Impang. [78]

When comparing the Tinguian versions of these fables with those of the Ilocano, one is impressed with the fact that while the incidents upon which they are founded are often identical, the stories themselves have frequently been moulded and changed by the tellers, who have introduced bits of old customs and beliefs until they reflect, in a way, the prevalent ideas of the people. Thus in the story of the magic poncho, [79] which is evidently of Spanish introduction, the owner is identified as the banbantay—a well-known minor spirit. Again, the first part of tale 85 is identical with that of the Ilocano, but ends with the parents of the groom preparing the things used in the pakálon—a very necessary part of the Tinguian marriage ceremony.

The footnotes have called attention to the many incidents which have their parallels in other districts. Reference to these shows that a large percentage are found in the islands toward the south. While recognizing that similarity of incidents does not necessarily mean identity of origin, we must still give full credit to the effects of borrowing, even over great distances. The easy communication along the coast during the past four hundred years and the contact with Spanish and Christianized officials and traders will readily explain the likeness of the tales in Division III to those held in distant islands, or even in Europe, but, as just noted, these are now undergoing change. Doubtless a similar inflow had been taking place, although at a slower rate, long before the Spaniards reached the Islands, and Tinguian mythology has grown up as the result of blending of native tales with those of other areas, the whole being worked over and reshaped until it fitted the social setting.

Previous writers—among them Ratzel and Graebner [80]—have sought to account for certain resemblances in culture, between Malaysia, Polynesia, and America, by historical connection. A part of our material—such as that of the blood-clot child (p. 125), [81] the rape of the maiden by the vine which carries her to the sky (p. 33), [82] the magic flight (p. 75), [83] and magic growth (p. 38) [84]—may seem to lend support to such a theory. These similarities are assuredly suggestive and interesting, but it appears to the writer that the material is too scanty and the folklore of intervening lands too little known to justify us in considering them as convincing proof of borrowing over such immense distances. [85]

GENERAL RESULTS

Our study has brought out certain general results. We have seen that Tinguian folklore has much in common with that of other tribes and lands.

While a part of this similarity is doubtless due to borrowing—a process which can still be seen at work—a considerable portion of the tales is probably of local and fairly recent origin, while the balance appears to be very old. These older tales are so intimately interwoven with the ceremonies, beliefs, and culture of this people that they may safely be considered as having been developed by them. They are doubtless much influenced by present day conditions, for each story teller must, even unconsciously, read into them some of his own experiences and the current beliefs of the tribe. At the same time these traditional accounts doubtless exercise a potent influence on the thoughts, beliefs, and actions of the people. In Tinguian society, where custom still holds undisputed sway, these well-known tales of past times must tend to cast into the same mould any new facts or experiences which come to them.

We believe that we are justified when we take the viewpoint of the Tinguian and consider "the stories of the first times" as essentially very old. How old it is impossible to state definitely, but a careful analysis of our material justifies us in believing that they reflect a time before the people possessed terraced rice fields, when domestic work animals were still unknown, and the horse had not yet been introduced into their land. That these are not recent events is attested by the great part they all now play in the ceremonial and economic life. It is evident that outside influences of great importance were introduced at a period later than the time when the Chinese first began to trade along the coasts of the Philippines for the prized jars, which play such an important rôle in the mythology, are not to be identified as those of native make but are ancient Chinese vessels dating back at least to the fourteenth and perhaps even to the tenth century. [86]

It is probable that the glass, porcelain, and agate beads, which are second only to the jars in importance, are exceedingly old. Many ancient specimens are still in use and are held for as fabulous prices as are those found among the interior tribes of Borneo. Nieuwenhuis has shown that the manufacture of beads had become a great industry in the middle ages, and had extended even to China and Japan, whence the products may have spread contemporaneously with the pottery. [87]

We have seen that, for the most part, the life, customs, and beliefs which appear in our reconstruction of "the first times" agrees closely with present conditions; certain things which seem formerly to have been of prime importance—such as the sending of a betel-nut covered with gold to invite guests to a festival or ceremony—appear to have their echo in present conditions. The betel-nut which played such a momentous part in the old times still holds its place in the rituals of the many ceremonies, although it is not now much used in daily life. The magic of to-day is less powerful than formerly, but is still a tremendous force. The communication of the ancient people with other members of the animate world, as well as with the inanimate and spiritual, and their metamorphosis into animals and the like, offers nothing strange or inconsistent to the people of to-day. They even now talk to jars, they converse with spirits who come to them through the bodies of their mediums, and people only recently deceased are known to have had the power of changing themselves, at will, into other forms.

In short, there is no sharp break between the mode of thought of to-day and that exhibited in the folklore. It is true that the tales give sanction to some things not in agreement with Tinguian usage—such, for instance, as the marriage of relatives, or the method of disposing of the dead—and it may be that we have here a remembrance of customs which long ago fell into disuse.

In a previous paper [88] the writer showed that there have been many migrations into Abra from the north, south, and west. A part of the emigrants have become thoroughly amalgamated with the Tinguian people and have doubtless introduced some part of their material culture and beliefs. This helps us to understand such conflicts as we have already noted in regard to the place held by thunder and lightning in the spirit world, as to the future abode of the spirits of the departed, as well as other discrepancies which the limits of this paper have prevented us from discussing.

It is not impossible that those customs of "the first times," which are at variance with those of to-day, may represent older ideas which have been swamped, or, on the other hand, the memory of the strange customs once practiced by the emigrants may have caused them to be attributed to the people of the tales.

Finally, we believe that a study of Tinguian mythology has shown us that we can gain a real knowledge of the past of a people through their folklore; that we can secure an insight into their mental life; and can learn something of the valuation they attach to certain of their activities and beliefs, which to us may seem at the surface trite and trivial.

ABSTRACTS

I

1

Two women are gathering greens when a vine wraps around one and carries her to the sky. She is placed near to spring, the sands of which are rare beads. Small house near by proves to be home of the sun. Woman hides until owner goes into sky to shine, then goes to house and prepares food. Breaks up fish stick and cooks it. It becomes fish. Single grain of rice cooked in pot the size of a "rooster's egg" becomes sufficient for her meal. Goes to sleep in house. Sun returns and sees house which appears to be burning. Investigates and finds appearance of flames comes from beautiful woman. Starts to prepare food, but awakens visitor. She vanishes. Each day sun finds food cooked for him. Gets big star to take his place in sky; returns home unexpectedly and surprises woman. They chew betel-nut together and tell their names. The quids turn to agate beads, showing them to be related, and thus suitable for marriage. Each night sun catches fish, but woman refuses it, and furnishes meat by cooking fish stick.

Woman decides to go with husband on daily journey through sky. When in middle of heavens she turns to oil. Husband puts her in a bottle and drops it to earth. Bottle falls in woman's own town, where she resumes old form and tells false tale of her absence. She becomes ill, asks mother to prick her little finger. Mother does so and child pops out. Child grows each time it is bathed. Girl refuses to divulge name of child's father. Parents decide to celebrate balaua and invite all people. Send out oiled betel-nuts covered with gold to invite guests. When one refuses, nut begins to grow on his knee or prized animal until invitation is accepted. Child is placed by gate of town in hopes it will recognize its father. Gives no sign until sun appears, then goes to it. Sun appears as round stone. Girl's parents are angry because of her choice of a husband and send her away without good clothes or ornaments.

Sun, wife and child return home. Sun assumes form of man. They celebrate balaua and invite all their relatives. Guests chew betel-nuts and the quid of the sun goes to that of Pagbokásan, so it is known that the latter is his father. Parents of sun pay marriage price to girl's people.

2

Aponibolinayen who is very ill expresses a desire for mangoes which belong to Algaba of Dalaga. Her brother dispatches two men with presents to secure them. One carries an earring, the other an egg. On way egg hatches and soon becomes a rooster which crows. They spread a belt on the water and ride across the river. When they bathe, the drops of water from their bodies turn to agate beads. Find way to Algaba's house by following the row of headbaskets, which reaches from the river to his dwelling. Defensive fence around the town is made up of boa constrictors, which sleep as they pass. Algaba seizes his spear and headaxe intending to kill the visitors, but weapons shed tears of oil. He takes other weapons, but they weep tears of blood. He then makes friends of the intruders. Learning their mission he refuses their gifts, but gets fruit and returns with them to their town. On way he uses magic and causes the death of Aponibolinayen. He takes her in his arms and restores her to life. While she rests in his arms, their rings exchange themselves. They chew betel-nuts and tell their names. The quids turn to agate beads and lie in rows. This is good sign. They marry and go to Algaba's town. They celebrate Sayang and send betel-nuts to invite their relatives. When the guests cross the river, the drops of water which run from their bodies are agate beads and stones of the river are of gold. Guests all chew betel-nut and lay down their quids. By arrangement of quids they learn the true parents of Algaba. His brother-in-law wishes to marry his new found sister and offers an engagement present. An earring is put in a jar and it is at once filled with gold, but Algaba lifts his eyebrows and half of the gold vanishes. Another earring is put in jar, and it is again full. Marriage price is paid later.

3

Aponitolau falls in love with girl he meets at the spring. They chew betel-nuts and tell their names. Girl gives false name and vanishes. Aponitolau sends his mother to arrange for his marriage with the girl. She wears a hat which is like a bird, and it gives her a bad sign, but she goes on. She crosses river by using her belt as a raft. The girl's parents agree to the match and price to be paid. Girl accepts a little jar and agate beads as engagement present. When Aponitolau goes to claim bride, he finds he is betrothed to wrong girl. His parents celebrate Sayang and invite many people, hoping to learn identity of girl at spring. She does not attend, but Aponitolau finds her among betel-nuts brought him by the spirit helpers. They chew betel-nuts and learn they are related and that both possess magical power.

After their marriage Aponitolau goes to his field. There he keeps many kinds of jars which act like cattle. He feeds them with lawed leaves and salt. While he is gone, the woman to whom he was first betrothed kills his new wife. He restores her to life. Takes her and her parents to the field to see him feed his jars.

4

A bird directs Aponitolau in his search for the maiden Asibowan. Girl furnishes him with food by cooking a fish stick. They have a daughter who grows one span each time she is bathed. Aponitolau discovers that his parents are searching for him, and determines to go home. Asibowan refuses to accompany him, but uses magic and transfers him and child to his town.

Aponitolau falls in love with girl he sees bathing, and his mother goes to consult her parents. She crosses river by using her belt as a raft; when she bathes, the drops of water from her body become agate beads. The girl's people agree to the marriage and accept payment for her.

Aponitolau and his bride celebrate Sayang and send out betel-nuts to invite the guests. Asibowan refuses to attend, but a betel-nut grows on her pig until, out of pity, she consents.

After the ceremony the brother of the bride turns himself into a firefly and follows her new sister-in-law. Later he again assumes human form and secures her as his wife.

5

The mother of Gawigawen is well received when she goes to seek a wife for her son. The girl's mother furnishes fish by breaking and cooking the fish stick. A day is set for payment of the marriage price. Guests assemble and dance. When bride dances she is so beautiful that sunshine vanishes, water from the river comes up into the town and fish bite her heels. When she arrives at her husband's home, she finds sands and grass of spring are made up of beads, and the walk and place to set jars are large plates. Her husband cuts off head of an old man and a new spring appears; his blood becomes beads and his body a great shade tree. Bride who has not yet seen the face of her husband is misled by evil tales of jealous women, and believes him to be a monster. During night she turns to oil, slips through floor and escapes. In jungle she meets rooster and monkey, who tell her she is mistaken and advise her to return home. She continues her way and finally reaches ocean. Is carried across by a carabao which at once informs its master of the girl's presence.

The master comes and meets girl. They chew betel-nut, and the quids turn to agate beads, so they marry.

They make Sayang and send betel-nuts to summon relatives. Nuts grow on pet pigs of those who refuse to go.

Guests are carried across river by betel-nuts. During dance Gawigawen recognizes his lost wife and seizes her. Is speared to death by the new husband, but is later brought back to life. In meantime the alan (spirits) inform the parents of the new groom that he is their child (from menstrual blood). Parents repay Gawigawen for his lost bride, and also make payment to the girl's family.

6

The enemies of Aponibolinayen, thinking her without the protection of a brother, go to fight her. She glances off their spears with her elbows. Her weapons kill all but Ginambo, who agrees to continue fight in one month.

Aponigawani has a similar experience with her enemies. A month later the two women meet as they go to continue the fight against their foes. They chew betel-nut, and quid of Aponibolinayen is covered with gold and that of her companion becomes an agate bead. They agree to aid each other. Go to fight and are hard pressed by foes. Spirit helpers go to summon aid of two men who turn out to be their brothers—were miscarriage children who had been raised by the alan. They go to aid sisters and kill so many people that pig troughs are floating in blood. One puts girls inside belt. They kill all the enemies and send their heads and plunder to the girls' homes. Brothers take girls to their parents. Father and mother of Aponigawani celebrate balaua and summon guests by means of oiled betel-nuts covered with gold. Guests chew betel-nut and spittle of children goes to that of parents, so relationship is established. Alan explain how they raised the miscarriage children. Heads of enemies are placed around the town and people dance for one month. Aponibolinayen marries brother of Aponigawani, who in turn marries the brother of her friend. Usual celebration and payments made. Relatives receive part of price paid for brides.

7

Aponitolau dons his best garments, takes his headaxe and spear, and goes to fight. When he reaches the spring which belongs to the ten-headed giant Giambólan, he kills all the girls, who are there getting water, and takes their heads. The giant in vain tries to injure him. Spear and headaxe of Aponitolau kill the giant and all the people of his town and cut off their heads. Heads are sent in order to hero's town—giants' heads first, then men's, and finally women's. On return journey Aponitolau is followed by enemies. He commands his flint and steel to become a high bank which prevents his foes from following. Upon his arrival home a great celebration is held; people dance, and skulls are placed around the town.

8

Aponitolau and his wife decide to celebrate Sayang, but he goes first to take the head of old man Ta-odan. He uses magic and arrives at once where foe lives. They fight and Ta-odan is beheaded. While Aponitolau is gone, an Ilocano comes to town and tries to visit his wife. She at first refuses to see him, but when he returns a needle she has dropped he puts a love charm on it. She then receives him into house. He remains until Aponitolau returns, then leaves so hastily he forgets his belt of gold. Woman hides belt in rice granary, but it reveals self by shining like fire. Aponitolau is suspicious and determines to find owner. As guests arrive for the celebration, he tries belt on each until he finds right one. He cuts off his head and it flies at once to his wife's breasts and hangs there. She flees with her children. They reach town, which is guarded by two kinds of lightning, but they are asleep and let them pass. They sleep in the balaua and are discovered by the owner of the place, who turns out to be an afterbirth brother of the woman. He removes the head of the dead Ilocano from her breasts. Betel-nuts are sent to summon their father and mother, who are surprised to learn of their afterbirth son. He returns home with them. Aponitolau fails to be reconciled to his faithless wife.

9

Ayo is hidden by her brother, but meets Dagdagalisit, who is fishing, and becomes pregnant. Child pops out between third and fourth fingers when Ayo has her hand pricked. Baby objects to first name; so is called Kanag. Milk from Ayo's breasts falls on her brother's legs while she is lousing him, and he thus learns of the child. He determines to build a balaua and invite all people, so he may learn who the father is. Sends out oiled betel-nuts to invite the guests and when one refuses to attend they grow on him or his pet pig. Dagdagalisit attends wearing only a clout of dried banana leaves. Brother of Ayo is enraged at her match and sends her and the baby away with her poor husband. When they arrive at her new home, Ayo finds her husband a handsome man who lives in a golden house, and whose spring has gravel of gold and agates. They summon their relatives to celebrate balaua with them. While Ayo's brother is dancing, her husband cuts off his head, but he is brought back to life. Ayo's husband pays her parents for her, but half the payment vanishes when her mother raises eyebrows. Husband again completes payment. They chew betel-nut and the quids of the children go to those of their parents. Dagdagalisit's parents learn he is a miscarriage child who was cared for by the alan (spirits).

10

Aponibalagen uses magic to create a residence in the ocean for his sister. Takes her and companions there on backs of crocodiles. Returns home.

Ingiwan who is walking is confronted by high bank and is forced to cross the ocean. Rides on his headaxe past the sleeping crocodiles which guard the maiden. Turns self into firefly and reaches girl. Assumes own form and chews betel-nut with her. Omens are good. He returns home and soon maiden is troubled with intense itching between her last fingers. She has place pricked, and baby boy pops out. Child grows one span at each bath. Aponibalagen learns of child when milk from sister's breasts falls on him. He takes her home and prepares to celebrate balaua. Oiled betel-nuts are sent to summon guests. They grow on knees of those who refuse to attend. Ingiwan, poorly clad, appears at the ceremony and is recognized by the child but not by its mother. Girl's brother, in rage, sends her away with the stranger. He assumes own form and proves to be handsome and wealthy. When they celebrate balaua, they chew betel-nut and thus learn who are his true parents.

11

When Aponitolau goes to visit his cousin, he finds him celebrating Sayang. He is incensed because no invitation has reached him, so sits in shade of tree near the spring instead of going up to the village. He finds the switch lost by Aponibolinayen. He is induced to attend the ceremony, where he meets with an old enemy, and they fight. The hawk sees the struggle and reports the death of Aponitolau to his sister. She sends her companions to avenge the death and they kill many people before they learn that the hawk was mistaken. Aponitolau restores the slain to life. He agrees to fight his enemies in two months. Before he goes to battle he summons the old men and women, and has them examine a pig's liver and gall. The omens are favorable. During the fight he becomes thirsty and his headaxe supplies him with water. He stops the slaughter of his enemies when they agree to pay him one hundred valuable jars. The jars and heads of the slain take themselves to his home. A celebration is held over the heads, and skulls are exhibited around the town.

Aponitolau goes to return the switch of Aponibolinayen. They chew betel-nuts and tell their names. Their finger rings exchange themselves, while their betel quids turn to agate beads and arrange themselves in lines—a sign of relationship. He cooks a stick and it becomes a fish. The girl vanishes, but Aponitolau turns himself into a firefly and finds her. They remain together one night, then he departs. On his way home he is seized by an immense bird which carries him to an island guarded by crocodiles. He is forced to marry a woman also captured by the bird.

Aponibolinayen gives birth to a child called Kanag. Child is delivered when an itching spot on mother's little finger is pricked. Kanag is kept in ignorance of father's fate until informed by an old woman whom he has angered. He goes in search of his father. By using power of the betel-nut he is enabled to cross the water on the backs of sleeping crocodiles. He kills gigantic snakes and finally the bird which had carried away his father. He takes father and the captive woman back home. Both women claim Aponitolau as husband. A test is held and Aponibolinayen wins.

12

Pregnant woman expresses desire for fruit of bolnay tree. Her husband asks what it is she wishes, and she falsely tells him fish roe. He uses magic to catch all fish in the river, and selects one with roe, releases others. She throws it to the dogs, and tells husband it is the liver of a deer she needs. He secures it, but when it likewise is fed to the dogs, he changes self into an ant and hides near wife until he learns her real wish. He secures the bolnay fruit, but upon his return allows his sweethearts to get all but a small piece of it. His wife eats the bit left and desires more. She quarrels with husband, who in rage drags her to the bolnay tree and places her in a hole. Her child Kanag is born when an itching spot between her third and fourth fingers is pricked. Child grows with each bath. He agrees to go with other boys to fight. Plants a lawed vine which is to keep his mother informed as to his condition. Child's father is with war party, but does not recognize son. It rains continually so party cannot cook; but the spirit helpers of child's mother feed him, and he shares food with companions. They plan ambush near enemies' town. Kanag cuts off head of a pretty girl; his companions kill an old man and woman. They return home and hold dance around the heads. When Kanag dances, earth trembles, coconuts fall, water from river enters the town, and the fish lap his feet. His father is jealous and cuts off his head. His mother sees lawed vine wilt and knows of son's death. Informs her husband he has killed son. She restores Kanag to life and they leave. Husband tries to follow, but magic growth of thorns in trail prevents. He is finally reconciled to his family and has former sweethearts killed.

13

A pregnant woman desires the fruit of an orange tree which belongs to the six-headed giant Gawigawen. Her husband asks her what it is she desires and she replies falsely; first, that she wishes a certain fruit, then fish roe, and finally deer liver. He secures each, taking the roe and liver out of the fish and deer without causing their death. Each of the articles makes the woman vomit, so her husband knows that she is not satisfied. Transforming self into a centipede he hides until he learns her real wish. Arms self and starts on perilous mission, but first plants lawed vine in house. By condition of vine wife is to know of his safety or death.

On way small dog bites him; he is tested by lightning and by thunder, and in each case gets a bad sign, but continues journey. Sails over ocean on his headaxe. Reaches cliff on which the town of the giant is placed, but is unable to scale it. Chief of spiders spins a web on which he climbs. Giant promises him the fruit provided he eats whole carabao. Chiefs of ants and flies calls their followers and eat animal for him. Is allowed to pick fruit, but branches of tree are sharp knives on which he is cut. He puts two of oranges on his spear and it flies away to his home. He dies and lawed vine at his house withers. Giant uses his skin to cover end of drum, puts his hair on roof of house and places his head at gate of town. Wife gives birth to child, which grows one span each time it is bathed. While still very small child angers old woman who tells him of his father's fate. Child determines to go in search of father despite mother's protests. On journey he meets all the tests put to his father, but always receives good signs. Jumps over cliff father had climbed on the spider web. He challenges giant to fight and shows valor by refusing to be the first to use his weapons. Giant unable to injure him, for he first becomes an ant, then vanishes. He throws his spear and it goes through giant, while his headaxe cuts off five of adversary's heads. Spares last head so it can tell him where to find his father. Collects father's body together and restores it to life. Lawed vine at their home revives. Father tries to cut off last head of giant, but fails; son succeeds easily. They send the headaxes to kill all people in town. Slaughter is so great the father swims in blood, but son stands on it. Both return home and hold a great celebration over the heads.

The father's spittle is lapped up by a frog which becomes pregnant. Frog gives birth to baby girl which is carried away by anitos. Girl is taught to make dawak (the duties of a medium). Her half brother hears her, changes self into a bird and visits her in the sky. Is hidden in a caldron to keep anitos from eating him. Tries to persuade sister to return with him. She promises to go when their father celebrates balaua. The ceremony is held and girl attends. Is so beautiful all young men try to obtain her. They are so persistent that brother returns her to sky where she still lives and aids women who make dawak.

14

Aponitolau and his wife plant sugar cane, and by use of magic cause it to grow rapidly. The daughter of the big star sees the cane and desires to chew it. She goes with her companions and steals some of the cane, which they chew in the field. Aponitolau hides near by and sees stars fall into the cane patch. He observes one take off her dress and become a beautiful woman. He sits on her garment and refuses to give it up until they chew betel-nut together. The star girl falls in love with him and compels him to return with her to the sky. Five months later she has a child which comes out from space between her last two fingers. Aponitolau persuades her to allow him to visit the earth. He fails to return at agreed time, and stars are sent to fetch him. He returns to the sky, but visits the earth again, eight months later. Earth wife bears him a child and they celebrate Sayang. Sky child attends and later marries an earth maiden.

15

The wife of Aponitolau refuses to comb his hair; so he has another woman do it. She, in turn, refuses to cut betel-nut for him to chew. While doing it for himself he is cut on his headaxe. The blood flows up into the air, and does not cease until he vanishes. Ceremonies made for him are without avail.

Aponitolau finds himself up in the air country. He meets maiden who is real cause of his plight. They live together and have a child which grows every time it is bathed. Aponitolau takes boy down to earth to visit his half brother. While there the tears of the mother above fall on her son and hurt him. They celebrate Sayang and the sky mother attends. After it is over the half brothers marry earth girls.

16

Ayo gives birth to three little pigs. Husband is ashamed, and while wife is at the spring he places the animals in a basket and hangs it in a tree. Basket is found by old woman, Alokotán, who takes it home. Pigs soon turn into boys. When grown they go to court the girls while they spin. Ayo hears of their visits and goes where they are. Milk from her breasts goes to their mouths and thus proves her to be their mother.

They celebrate balaua. Ayo puts one grain of rice in each of twelve jars and they are at once filled with rice. Betel-nuts summon the people to attend the ceremony. The old woman Alokotán attends and the whole story of the children's birth and change to human form comes out.

17

Dumalawi makes love to his father's concubines who openly show their preference for the son. The father plans to do away with the youth. Gets him drunk and has storm carry him away. Dumalawi awakens in center of a large field. He causes betel trees to grow, then cuts the nuts into bits and scatters them on the ground. The pieces of nut become people who are his neighbors. He falls in love with daughter of one of these people and marries her. They celebrate Sayang and send out oiled betel-nuts to invite the guests. All guests, except Dumalawi's father, are carried across river on the back of a crocodile. Animal at first dives and refuses to carry him, but finally does so. All drink from a small jar which still remains a third full. Parents of Dumalawi pay the usual marriage price for girl, but her mother insists on more. Has spider spin web around the town, and groom's mother has to cover it with golden beads.

18

While two women are bathing, blood from their bodies is carried down stream. Two alan secure the drops of blood and place them in dishes. Each drop turns into a baby boy. Boys go to fight and kill many people at the spring. They challenge a ten-headed giant. He is unable to injure them, but their weapons kill him and his neighbors. Heads of the victors take themselves to homes of the boys. A storm transports the giant's house. Boys trample on town of the enemy and it becomes like the ocean. They use magic and reach home in an instant. Hold celebration over the heads. Some guests bring beautiful girls hidden in their belts. Alan tell history of lads and restore them to their people. One of boys falls in love and his parents negotiate match for him. The payment for the girl is valuable things sufficient to fill balaua eighteen times, and other gifts in her new home.

19

Kanag is lead by his hunting dog to a small house in the jungle. Girl who lives there hides, but appears on second day. They chew betel-nuts and tell their names. The quids turn to agate beads and lie in order, showing them to be related and hence suitable for marriage. They remain in forest two years and have children. Kanag uses magical power and transfers their house to his home town during night. Children see sugar cane which they wish to chew. Kanag goes to secure it, and while away his mother visits his wife and abuses her. She becomes ill and dies. Kanag tries to kill his mother, but fails. Puts body of wife on a golden raft, places golden rooster on it and sets afloat on the river. Rooster crows and proclaims ownership whenever raft passes a village. Old woman Alokotán secures raft before it vanishes into the hole where river ends. Revives the girl. Kanag and children reach home of Alokotán, and girl is restored to them. They celebrate balaua and send betel-nuts covered with gold to invite relatives. When guests arrive, they chew betel-nut and learn that Kanag and his wife are cousins. Kanag's parents pay marriage price, which is the balaua filled nine times with jars. Girl's mother raises eyebrows and half of jars vanish. Balaua is again filled. Guests dance and feast. Part of marriage price given to guests.

20

Kanaa's sweetheart desires the perfume of Baliwán and promises to fulfill his desires if he secures it for her. Gives him arm beads from left arm in token of her sincerity.

Kanag and a companion set out on mission but are warned, first by a jar and later by a frog, not to continue. They disregard the advice and go on. They reach the tree on which perfume grows, and Kanag climbs up and breaks off a branch. He turns into a great snake, and his companion flees. Snake appears to Langa-ayan and proves its identity by the arm beads around its neck. She takes it to a magic well, the waters of which cause the snake skin to peel off, and the boy is restored to his own form. Kanag marries Amau, and when they celebrate balaua he returns the bracelet to his former sweetheart. His parents fill the balaua nine times with valuable articles, in payment for his bride.

21

Kanag is sent to watch the mountain rice, although it is well protected from wild pigs. Thinks parents do not care for him, is despondent. Changes self into an omen bird and accompanies his father when he goes to fight. Father obeys signs and secures many heads from his enemies. He holds a great celebration over the heads, but Kanag refuses to attend. Decides to go down to earth to eat certain fruits. Parents order their spirit helpers to accompany him and dissuade him if possible. They show him a beautiful girl with whom he falls in love. He assumes human form and meets her. They chew betel-nut and tell their names. Signs are favorable for their marriage. His parents agree to fill the balaua nine times with various kinds of jars. They do so, but mother of girl raises eyebrows and half of jars vanish and have to be replaced. Girl's mother demands that golden beads be strung on a spider web which surrounds the town. This is done, but web does not break. Girl's mother hangs on thread which still holds. She then agrees to the marriage. Guests dance and then return home, each carrying some of the jars.

22

While Ligi is bathing in river his headband flies away and alights on the skirt of a maiden who is bathing further down stream. The girl carries the headband home and soon finds herself pregnant. The child is born when she has the space between her third and fourth fingers pricked. With each bath the child grows a span and soon becomes so active that he hinders mother at her work. She decides to put him with his father during daytime. Uses magic and causes people of the town to sleep while she places child beside father. Ligi awakes and finds child and his headband beside him. Child refuses to answer questions. Mother secures child at nightfall and repeats acts next day. Child is hidden, so she fails to get him. Ligi determines to learn who mother of child is; sends out oiled betel-nuts covered with gold to invite all people to a Sayang. When summoned, the mother refuses to go until a betel-nut grows on her knee and compels her. She goes disguised as a Negrito, but is recognized by the child who nurses from her while she is drunk. Ligi suspects her, and with a knife cuts off her black skin. Learns she is child's mother and marries her. He divorces his wife Aponibolinayen, who marries husband of Gimbagonan. The latter poisons her rival, but later restores her, when threatened by her husband.

23

A flock of birds offer to cut rice for Ligi. He agrees, and goes home with a headache. Birds use magic so that the rice cutters work alone, and the tying bands tie themselves around the bundles. The birds each take one grain of rice in payment. They use magic again so that bundles of rice take themselves to the town. Ligi invites them to a ceremony, and then follows them home. He sees them remove their feathers and become one girl. They go back to the celebration, where all chew betel-nut. Girl's quid goes to those of her parents, from whom she had been stolen by the spirit Kaboniyan. The parents of Ligi pay the usual marriage price for the girl.

24

When the husband of Dolimáman pricks an itching spot between her third and fourth fingers, a baby boy pops out. Child who is called Kanag grows each time he is bathed. While his wife is away the father puts child on a raft and sets it afloat on the river. Child is rescued by old woman Alokotán, who is making a pool in which sick and dead are restored to health. Boy plays on nose flute which tells him about his mother, but he does not understand. Plays on bunkaka with same result. Mother who is searching her child passes by while he is playing. Milk from her breasts goes to his mouth, and she recognizes him. They stay with old woman despite pleading of husband.

25

Awig sends his daughter to watch the mountain rice. She stays in a high watch house, but is found by tattooed Igorot, who cut her body in two and take her head. Father goes to seek her murderers, but first plants a lawed vine in the house; by its condition his wife is to know of his safety or death. He climbs high tree and looks in all directions. Sees Igorot, who are dancing around the head of his daughter. He takes juice from the poison tree and goes to the dance, where he is mistaken for a companion. He serves liquor to others and poisons them. Takes daughter's head and starts home. Is followed by four enemies. Uses magic and causes cogon field to burn, so foes are delayed. Repeats this several times and finally escapes. He joins head and body of his daughter, and old woman Alokotán puts saliva on cuts and revives her. Old woman places four sticks in the ground and they become a balaua. Betel-nuts are sent out to invite guests and many come. When the girl dances with her lover, the water comes up knee deep into the town and they have to stop. She is engaged and her lover's parents fill the balaua three times with valuable gifts, in payment for her. Half of gifts vanish, when her mother raises her eyebrows, and are replaced.

Her husband discovers the scar on her body where Igorot had cut her. Takes her to magic well where she bathes. Scars vanish.

26

The mother of Dumanágan negotiates marriage for her son with Aponibolinayen. Brother of girl puts her in his belt and carries her to place where agreement is made. When they reach gate of town, young girls offer them cakes, in order to take away bad signs seen on road. Boy's parents pay for girl and they marry. She gives birth to son named Asbinan. He marries Asigowan, but his jealous concubines cause her to cut her finger and she dies. Her body is placed in a tabalang on which a rooster sits, and is set afloat on the river. Crowing of the cock causes old woman Alokotán to rescue the corpse. She places it in her magic well and the girl is again alive and beautiful. She returns to her husband as a bird; is caught by him and then resumes own form.

27

Baby of four months hears his father tell of his youthful exploits. Decides to go on head hunt despite protests of parents. Is detained on his trip by young alan girls. Finally reaches Igorot town and by means of magic kills all the people and takes their heads. Heads take themselves to his home. On way back he plays bamboo jew's harp and it summons his brothers to come and see him. They chew betel-nut and make sure of relationship. Continuing his journey, he is twice lost. Finds an unknown sister hiding among lawed vines. Puts her in his belt and carries her home. Upon his arrival a celebration is held and the new found brothers and sister, who had been stolen by alan, are restored to parents.

28

The mother and caretaker of Asbinan try to arrange for him to marry Dawinisan, but are refused. Asbinan goes to the girl's home and feigns sickness. Is cared for by the girl, who becomes infatuated with him and accepts his suit. His parents pay jars and gold—in the shape of deer—for her.

29

Asbinan refuses to eat until his father secures fish roe. He then demands Chinese dishes from the coast town of Vigan. When these are supplied, he eats, and then demands the love charm which his father used when a young man. He goes to the place where the maidens are spinning, and when one offers to give him a light for his pipe, he blows smoke in her face. The charm acts and she becomes ill. He convinces her people that the only way she can be cured is by marrying him. Her parents accept payment for the girl.

30

Tolagan decides to visit certain places in Pangasinan. He rides on a pinto pony and carries rice cakes as provisions. At the spring in Kaodanan he meets a beautiful maiden who warns him to return home, because the birds have given him a bad sign. He returns only to find that his wife has been stolen by the spirit Kaboniyan. He fails to find her, but is comforted by winning a new bride (probably the girl of Kaodanan).

31

Two girls are adopted by a rich man, who treats them as his daughters, except that he does not offer them bracelets or rings. They dress as men and go to see a jeweler. Two young men suspect and follow them, but they succeed in escaping and return home.

The spirit helpers of the youths take the forms of hawks and finally locate the maidens, whom they carry away. The youths plan to marry the girls and invite many friends to the celebration. Kanag and his companion attend, become enamored with the brides and steal them. Upon chewing betel-nuts they learn that they are related, so they are married.

II

32

The Ipogau who are trying to celebrate Sayang make errors. The spirit Kadaklan and his wife instruct them to go and watch the Sayang at Sayau. They do as bidden and after learning all the details return home and perform the ceremony. The chief spirits are pleased and cause the lesser spirits to attend the ceremony when summoned by the medium. The sick improve.

33

The people who are conducting the Dawak ceremony fail to do it properly. Kaboniyan (a spirit) goes down and instructs them. After that they are able to cure the sick.

34

The spirits of Dadaya notice that their feather headdresses have lost their lustre. They place them on the house of some mortals, who at once become ill. The spirit Kaboniyan instructs them to make the Pala-an ceremony. They obey, the feathers regain their brightness and the people recover.

35

The father who is starting for a head-dance agrees to meet his wife and baby at sun down. When he reaches the agreed spot, he finds only their hats; he looks down and sees them in the ground. He tries in vain to get them out. The spirit Kaboniyan instructs him to perform the Ibal ceremony. He does so and receives his wife and child.

36

The spirit Ináwen, who lives in the sea, sends her servants to spread sickness. They kill many people who fail to make the Sangásang ceremony. A man is disturbed at night by barking of dogs, goes to door and meets a big spirit which has nine heads. Spirit tells him how to make the offering in Sangásang. He follows directions and spirits carry gift to their mistress. She mistakes the blood of a rooster for that of human beings. Is displeased with the taste and orders spirits to stop killing.

37

The spirit Maganáwan sends his servants to secure the blood of a rooster mixed with rice. People see many snakes and birds near gate of town. They make the ceremony Sangásang and offer blood and rice. The servants of Maganáwan carry the offering to him. He takes it in his mouth and spits it out, and in the same way the sickness is removed from the mortals.

38

The people who are digging holes for house poles get a bad sign from the omen bird. They abandon the place and dig again. The deer gives a bad sign, then the snake, then different birds. They change locations many times, but at last ignore the signs and complete the house. The family are continually in trouble and are ill.

The spirit Kaboniyan goes to see the sick persons; he lets his spear drop through the house, and then tells them the cause of the trouble is that they have failed to make Sangásang. He instructs them what to do, and when they obey all become well.

39

The different parts of the house quarrel and each insists on its importance. At last they recognize how necessary each one is for the other and cease their wrangling; then the people who live in the house are again in good health.

40

The great spirit sees the people of Bisau celebrating the Ubaya ceremony, and determines to reward them by increasing their worldly goods. He appears as a man and rewards them.

41

Dayapán, who has been ill for seven years, goes to bathe. The spirit Kaboniyan enters her body and instructs her how to perform healing ceremonies. He also teaches her how to plant and reap, and she in turn teaches the Tinguian. While she is bathing she ties a cock and dog by the water side. The dog eats the cock, and thus death comes into the world.

42

Girl who lacks certain organs is ashamed to marry. She is sent by her mother to cause lameness to people who pass. A man who falls victim to her magic is only cured when the girl instructs him how to make the Bawi ceremony.

43

The spirit Kaboniyan instructs a sick man to make offerings at the guardian stones. He does as bidden and becomes well. They perform ceremonies near the stones when they go to fight or celebrate balaua, and sometimes the spirit of the stones appears as a wild rooster, a white cock, or a white dog. A man who defiles the stones becomes crazy.

44

Man sees a woman walking at night near the guardian stones. She refuses to talk and he cuts her in the thigh. She vanishes into the stones. Next day it is seen that one of the stones is cut. Man dies.

45

The old men of Lagayan see peculiarly shaped stones traveling down the river, accompanied by a band of blackbirds. They catch the stones and carry them to the gate of the village, where they have since remained as guardians.

46

The spirit Ibwa visits a funeral and is given some of the juices, coming from the dead body, to drink. Since then he always tries to eat the body of the dead unless prevented. He is accompanied by another evil spirit whose embrace causes the living to die.

47

A widow leaves the town before the period of mourning for her husband is past. The spirit appears first to the daughter-in-law and is fed by her, then asks for his wife. He goes to the place where she is watching the corn and sleeps with her. She apparently becomes pregnant, but fails to be delivered, and dies.

48

Two men agree to hunt carabao the following morning. In the night one dies, but the other not knowing this leaves the town and goes to the appointed place. He meets the spirit of the dead man, and only saves his life by running his horse all the way home.

49

A man and his wife are living near to their field when the husband dies. An evil spirit comes to the door, but is driven away by the wife w with a headaxe. Several evil spirits attempt to gain entrance; then the chief comes. He breaks down the door; he cuts off the dead man's ears and makes the woman chew them with him—like betel-nut. The signs are propitious. He changes the woman's two breasts into one, in the center of her chest, and takes her home.

50

A man, whose brother has just died, goes to hunt. He begins to cut up the game when his brother's spirit appears. He feeds it, but food comes out of its anus as fast as it eats. He flees and is pursued by the spirit until, by chance, he runs among alangtin bushes. The spirit dislikes the bush and leaves.

51

The people fail to put the banal vine and iron on the grave. An evil spirit notices the omission and steals the body.

52

A man goes to hunt his carabao in the mountains. He fails to plant branches at his head before he sleeps. A spirit expectorates on him, and he soon dies.

53

Two men who have to sleep in the mountains make beds of sobosob leaves. In the night they hear the evil spirits come and express a desire to get them. Spirits dislike the leaves, so do not molest the men.

54

Three hunters spend the night in the open. One covers himself with a red and yellow striped blanket. In the night two spirits come and think he is a little wild pig, and decide to eat him. The hunter hears them and exchanges blankets with one of his companions. The companion is eaten, and hence the kambaya, or striped blanket, is no longer used on the trail.

55

The spirit Bayon steals a beautiful girl and carries her to the sky, where he changes her breasts into one and marries her. She drops her rice pounder to the earth, and thus her people learn of her fate. Both she and her husband still attend certain ceremonies.

56

A hunter is carried away by a great bird. He is placed in the nest with its young and aids in feeding them. When they are large, he holds on to them, and jumps safely to the ground. He goes to fight against his enemies. While he is gone his wife dies. Upon his return he sees her spirit driving a cow and two pigs. He follows her to the spirit's town and is hidden in a rice bin. When spirits try to get him during the night, he repels them by throwing feathers. Feathers become exhausted, and he is forced to return home.

57

A man encounters a large being, which, from its odor, he recognizes as the spirit of a dead man. He runs to get his friends, and they find the spot trampled like a carabao wallow.

58

The dead wife of Baluga harvests his rice during the nighttime. He hides and captures her. They go together to the spirit town, in the ground, and secure her spirit which is kept in a green bamboo cup. As they are returning to the ground they are pursued, but Baluga cuts the vine on which their pursuers are climbing. When they reach home, they hold a great celebration.

59

An alan takes the afterbirth and causes it to become a real child named Sayen. Afterbirth child marries a servant, thinking he has married her mistress. Learns he is deceived, and causes death of his wife; then kills many people in the town of the girl who has deceived him. She gets him to desist, and after he revives some of the slain marries him. People of neighboring town are troubled by the komau, an evil spirit, who always causes the death of as many people as the hunters have secured deer. Sayen kills the komau. He fights with the great spirit Kaboniyan. Neither is able to overcome the other, so they become friends. They fight together against their enemies. Sayen often changes himself into a fish or chicken, and hides after a fight. This is observed by people who set a trap and capture him. He is killed.

60

A man while in the woods hears the alan near him. He feigns death and the spirits weep for him. They put gold and beads on the body. He springs up and seizes the offerings. They demand the return of one bead; he refuses, and the spirits burn his house.

61

Two men who have killed a wild pig desire fire. One goes to house of an alan and tries to secure it while the spirit sleeps. She awakes and goes with the man to the pig. Man carries liver of the animal back to the baby alan. He eats the liver and then throws the child into a caldron of hot water. He tells his companion what he has done, and they climb a tree near the water. The alan discovers their hiding place by seeing their reflection in the water. She climbs up, feet first, but they cut the vine on which she is ascending, and she is killed. They go to her house and secure a jar of beads and a jar of gold.

62

The flat earth is made by the spirit Kadaklan. He also makes the moon and sun, which chase each other through the sky. The moon sometimes nearly catches the sun, but becomes weary too soon. The stars are stones, the lightning a dog.

63

A flood covers the land. Fire has no place to go, so enters bamboo, stones and iron. It still lives there and can be driven out by those who know how.

64

A man finds his rice field disturbed even though well fenced in. He hides and in middle of night sees some big animals fly into it. He seizes one and cuts off its wings. The animal turns out to be a mare which is pregnant and soon has male offspring. The place where the wings once grew are still to be seen on the legs of all horses.

65

A lazy man, who is planting corn, constantly leans on his planting stick. It becomes a tail and he turns into a monkey.

66

A boy is too lazy to strip sugar cane for himself. His mother in anger tells him to stick it up his anus. He does so and becomes a monkey.

67

A lazy girl pretends she does not know how to spin. Her companions, in disgust, tell her to stick the spinning stick up her anus. She does so and at once changes into a monkey.

68

A war party are unable to cross a swollen river. They wish to become birds. Their wish is granted and they are changed to kalau, but they are not able to resume the human forms. Those who wore the white mourning bands, now have white heads.

69

A mother puts a basket over her lazy son. When she raises it a bird flies away crying "sigakók" (lazy).

70

A young man who owns a rice field gets a new wife. He leaves her to harvest the crop. She is discouraged over the prospect and wishes to become a bird. Her wish is fulfilled, and she becomes a kakok.

71

The dog of Ganoway chases a deer into a cave. The hunter follows and in the darkness brushes against shrubs which tinkle. He breaks off some branches. Cave opens again on the river bank, and he finds his dog and the dead deer at the entrance. He sees that fruits on the branches he carries are agate beads. Returns, but fails to find more. His townspeople go with him to seek the wonderful tree, but part of the cave is closed by the spirit Kaboniyan who owns it.

72

The jar Magsawi formerly talked softly, but now is cracked and cannot be understood. In the first times the dogs of some hunters chased the jar and the men followed, thinking it to be a deer. The jar eluded them until a voice from the sky informed the pursuers how it might be caught. The blood of a pig was offered, as the voice directed, and the jar was captured.

73

The sun and moon fight. Sun throws sand in moon's face and makes the dark spots which are still visible.

74

A man who went with a war party is away so long that he does not recognize his daughter when he returns. He embraces her when she meets him at the town gate. In shame she changes herself into a coconut tree.

75

Two flying snakes once guarded the gap in the mountains by which the Abra river reaches the sea. Two brave men attack them with banana trunks. Their wings stick in the banana trees and they are easily killed. The men are rewarded with gold made in the shape of deer and horses.

76

A man named Tagápen, of Ilocos Norte, with his wife and child goes up the Abra river on a raft. They stop at various towns and Tagápen goes up to each while his wife comforts the child. They finally reached Patok where they go to live in the balaua. They remain there teaching the people many songs.

III

77

A turtle and a monkey go to plant bananas. The turtle places his in the ground, but the monkey hangs his in a tree. Soon the tree of the turtle has ripe fruit, but the monkey has none. Turtle asks monkey to climb and secure the fruit. Monkey eats all but one banana, then sleeps in the tree. Turtle plants sharp shells around the tree and then frightens monkey which falls and is killed. Turtle sells his flesh to other monkey and then chides them because they eat their kind. Monkeys catch turtle and threaten first to cut and then to burn him. He deceives them by showing them marks on his body. They tie weight to him and throw him into the water. He reappears with a fish. Monkeys try to imitate him and are drowned.

78

A turtle and lizard go to steal ginger. The lizard talks so loudly he attracts the attention of the owner. The turtle hides, but the lizard runs and is pursued by the man. The turtle enters the house and hides under a coconut shell. When the man sits on the shell the turtle calls. He cannot discover source of noise and thinks it comes from his testicles. He strikes these with a stone and dies. The turtle and the lizard see a bees' nest. The lizard hastens to get it and is stung. They see a bird snare and turtle claims it as the necklace of his father. Lizard runs to get it but is caught and killed.

79

A little bird calls many times for a boy to catch it. He snares it and places it in a jar. Lad's grandmother eats the bird. He discovers the theft, leaves home and gets a big stone to swallow him. The grandmother gets horses to kick the stone, carabao to hook it, and chickens to peck it, but without result. When thunder and her friends also fail, she goes home without her grandson.

80

A frog, which is attached to a hook, lures a fish so that it is caught.

81

The five fingers are brothers. The thumb goes to get bamboo. He tries to kiss the bamboo and his nose sticks. One by one the others go in search of the missing but are captured in the same manner. The little finger, which alone remains free, releases the others.

82

A carabao and a shell agree to race along the river. The carabao runs swiftly, then pauses to call "shell." Another shell replies and the carabao continues running. This is repeated many times until at last the carabao falls dead.

83

A crab and a shell go to get wood. The crab pulls the rope on his load so tightly that he breaks his big legs and dies. The shell finds his friend dead and cries until he belches his own body out of the shell and he dies.

84

A mosquito tells a man he would eat him were it not for his ears.

85

A messenger goes to negotiate a marriage. When he arrives he sees the people nodding their heads as they suck meat out of shells. He returns home without stating his mission, but reports an acceptance. Girl's people are surprised when people come for pakálon.

86

A man sees people eating bamboo shoots, and is told they are eating pagaldanen. He understands them to say aldan—"ladder," so he goes home and cooks his bamboo ladder. Is ridiculed by his friends.

87

A man with heavily laden horse asks the length of a certain trip. Boy replies, "If you go slowly, very soon; if you go fast, all day." The man hurries so that coconuts keep falling off the load and have to be replaced. It is dark when he arrives.

88

A woman eats the fruit belonging to crocodile and throws away the rind. Crocodile sees her tooth marks and recognizes the offender. He demands that she be given him to eat. Her people agree, but first feed him a hot iron. He swallows it and dies.

89

A lazy man goes to cut bamboo, and a cat steals his cooked rice. He catches the cat in a trap and takes it home. It becomes a fighting cock. The man starts for a cock fight, and on the way is joined by a crocodile, a deer, a mound of earth and a monkey. The rooster kills all the other birds at the fight, then the crocodile wins a diving contest, the deer a race, the mound of earth a wrestling match, and the monkey excels all in climbing. The man wins much money in wagers and buys a good house.

90

A spirit lets a man take his poncho which makes him invisible. He goes to his wife who recognizes his voice and thinks him dead. He takes off poncho and appears before her.

91

A fisherman is seized by a big bird which carries him to its nest. The small birds try to eat him, but he seizes one in each hand and jumps from the tree. He reaches the ground unhurt and returns home.

VITA

Fay-Cooper Cole

Born Plainwell, Michigan, August 8, 1881. Educated at University of Southern California, Northwestern University, Chicago University, Berlin University, Columbia University. B.S. Northwestern University, 1903.

Publications:

The Tinguian. Philippine Journal of Science, Vol. III, No. 4. 1908.

Distribution of the Non-Christian Tribes of Northwestern Luzon. Am. Anthro., Vol. II, No. 3. 1909.

The Bagobo of Davao Gulf. Philippine Journal of Science, Vol. VI, No. 3. 1911.

Chinese Pottery in the Philippines. Pub. Field Mus. Nat. Hist., Vol. XII, No. 1. Chicago, 1912.

Wild Tribes of Davao District, Mindanao. Pub. Field Mus. Nat. Hist., Vol. XII, No. 2. Chicago, 1913.

Traditions of the Tinguian. Pub. Field Mus. Nat. Hist., Vol. XIV, No. 1. Chicago, 1915.

NOTES

[1] Traditions of the Tinguian. (Pub. Field Museum of Natural History. Anthro. Series, Vol. No. I. Chicago, 1915.)

[2] Men or women through whom the superior beings talk to mortals. During ceremonies the spirits possess their bodies and govern their language and actions. When not engaged in their calling, the mediums take part in the daily activities of the village.

[3] See page 26.

[4] The initial portion of some of these names is derived from the respectful term apo—"sir," and the attributive copulate ni; thus the original form of Aponitolau probably was Apo ni Tolau, literally "Sir, who is Tolau." However, the storytellers do not now appear to divide the names into their component parts, and they frequently corrected the writer when he did so, for this reason such names appear in the text as single words. Following this explanation it is possible that the name Aponibolinayen may be derived from Apo ni bolan yan, literally "Sir (mistress) who is place where the moon"; but bolan generally refers to the space of time between the phases of the moon rather than to the moon itself. The proper term for moon is sinag, which we have seen is the mother of Gaygayóma—a star,— and is clearly differentiated from Aponibolinayen.

[5] [male]—male. [female]—female.

[6] Occasionally the storytellers become confused and give Pagbokásan as the father of Aponitolau.

[7] The town of Natpangán is several times mentioned as though it was the same as Kaodanan.

[8] The figures in parentheses refer to pages in the volume Traditions of the Tinguian, Pub. Field Mus. Nat. Hist., Vol. XIV, No. I. Chicago, 1915.

[9] The only possible exception to this statement is the mention of a carabao sled on p. 150, and of Aponitolau and Aponibolinayen riding on a carabao p.51. Traditions of the Tinguian. (Pub. Field Museum, vol. xiv, No. I; Chicago, 1915.)

[10] A term applied to any of the wilder head-hunting tribes.

[11] Ladders are placed on each side of the town gate and are inclined toward one another until they meet at the top. Returning warriors enter the

village by climbing up the one and descending the other, never through the gate.

[12] Copper gongs.

[13] Sharpened bamboo poles which pass through the foramen magnum.

[14] This poison is placed in the food or drink. The use of poisoned darts or arrows seems never to have been known to this people.

[15] A similar custom is found among the Kayan of Borneo. See Hose and McDougall, Pagan Tribes of Borneo, Vol. II, p. 171 (London, 1912).

[16] In this dance a man and a woman enter the circle, each holding a cloth. Keeping time to the music, they approach each other with almost imperceptible movements of feet and toes, and a bending at the knees, meanwhile changing the position of the cloths. This is varied from time to time by a few quick, high steps. For fuller description see article by author in Philippine Journal of Science, Vol. III, No. 4, 1908, p. 208.

[17] The custom was formerly practised by the Ilocano. See Reves, Folklore Filipino, p. 126 (Manila, 1899).

[18] See Philippine Journal of Science, Vol. III, No. 4, 1908, pp. 206, ff.

[19] The Tinguian do not have a classificatory system of relationship terms. The term kasinsiu is applied alike to the children of mother's and father's brothers and sisters.

[20] A sacred dance in which a number of men and women take part. It takes place only at night and is accompanied by the singing of the participants.

[21] The night preceding the greatest day of the Sayang ceremony.

[22] Runo, a reed.

[23] See p. 8, note 2.

[24] A short ceremony held for the cure of fever and minor ills. It also forms a part of the more extensive rites.

[25] A sugar-cane rum.

[26] See p. 7, note 1.

[27] Lesser spirits.

[28] Lesser spirits.

[29] Lesser spirits.

[30] Like ideas occur in the folk-tales of British North Borneo. See Evans, Journal Royal Anthro. Inst., Vol. XLIII, 1913, p. 444.

[31] In various guises the same conception is found in Europe, Asia, Africa, and Malaysia. See Cox, An Introduction to Folklore, p. 121 (London, 1904).—In an Igorot tale the owner captures and marries the star maiden, who is stealing his rice. Seidenadel, The Language of the Bontoc Igorot, p. 491 ff. (Chicago, 1909).

[32] The Dusun of Borneo have tales of talking jars. Evans, Journal Royal Anthro. Inst., Vol. XLIII, 1913, pp. 426-427. See also Cole and Laufer, Chinese Pottery in the Philippines (Pub. Field Museum of Nat. Hist., Vol. XII, No. I, p. 11 ff., 1912).

[33] Piper sp.

[34] Bagobo tales relate that in the beginning plants, animals, and rocks could talk with mortals. See Benedict, Journal American Folklore, Vol. XXVI, 1913, p. 21.

[35] Tales of animals who assist mortals are found in all lands; perhaps the best known to European readers is that of the ants which sorted the grain for Cinderella. See also Evans, Jour. Royal Anthro. Inst., Vol. XLIII, 1913, p. 467, for Borneo; Tawney's Kathá Sarit Ságara, pp. 361 ff., Calcutta, 1880, for India.

[36] Fabulous birds of gigantic size, often known under the Indian term garuda, play an important part in the beliefs of the Peninsular Malays.

[37] A similiar incident is cited by Bezemer (Volksdichtung aus Indonesien). See also the Bagobo tale of the Kingfisher (Benedict, Jour. American Folklore, Vol. XXVI, 1913, p. 53).

[38] The magic flight has been encountered in the most widely separated parts of the globe, as, for instance, India and America. See Tawney, Kathá Sarit Ságara, pp. 361, 367 ff. and notes, (Calcutta, 1880); Waterman, Jour. American Folklore, Vol. XXVII, 1914, p. 46; Reinhold Köhler, Kleinere Schriften, Vol. I, pp. 171, 388.

[39] In the Dayak legend of Limbang, a tree springs from the head of a dead giant; its flowers turn to beads; its leaves to cloth; the ripe fruit to jars. See H. Ling Roth, The Natives of Sarawak and British North Borneo, Vol. I, p. 372.

[40] Similar incidents are to be found among the Ilocano and Igorot in Borneo; in Java and India. See Reyes, Folklore Filipino, p. 34, (Manila, 1889); Jenks, The Bontoc Igorot, p. 202, (Manila, 1905); Seidenadel, The Language of the Bontoc Igorot. p. 491, 541, ff, (Chicago, 1909); Evans,

Journal Royal Anthro. Inst., Vol. XLIII, 1913, p. 462; Ling Roth, Natives of Sarawak and British North Borneo, Vol. I, p. 319; Tawney, Kathá Sarit Ságara, Vol. II, p. 3, (Calcutta, 1880); Bezemer, Volksdichtung aus Indonesien, p. 49, (Haag, 1904).

[41] This peculiar expression while frequently used is not fully understood by the story tellers who in place of the word "whip" occasionally use "make." In one text which describes the Sayang ceremony, I find the following sentence, which may help us to understand the foregoing: "We go to make perfume at the edge of the town, and the things which we take, which are our perfume, are the leaves of trees and some others; it is the perfume for the people, which we give to them, which we go to break off the trees at the edge of the town." Again in tale 20, Kanag breaks the perfume of Baliwán off a tree.—The use of sweetly scented oil, in raising the dead, is found in Dayak legends. See Ling Roth, The Natives of Sarawak and British North Borneo, Vol. I, p. 314.

[42] According to a Jakun legend, the first children were produced out of the calves of their mothers' legs. Skeat and Bladgen, Pagan Races of the Malay Peninsula, Vol. II, p. 185.—A creation tale from Mangaia relates that the boy Rongo came from a boil on his mother's arm when it was pressed. Gill, Myths and Songs of the South Pacific, p. 10 (London, 1876).

[43] This power of transforming themselves into animals and the like is a common possession among the heroes of Dayak and Malay tales. See Ling Roth, The Natives of Sarawak and British North Borneo, Vol. I, p. 312; Perham, Journal Straits Branch R., Asiatic Society, No. 16, 1886; Wilkinson, Malay Beliefs, pp. 32, 59 (London, 1906).

[44] The present day Tinguian attach much importance to these omens. The gall and liver of the slaughtered animal are carefully examined. If the fluid in the gall sack is exceedingly bitter, the inquirer is certain to be successful; if it is mild he had best defer his project. Certain lines and spots found on the liver foretell disaster, while a normal organ assures success. See also Hose and McDougall, Pagan Tribes of Borneo, Vol. II, p. 60 ff.

[45] See p. 21, note 1.

[46] The present capital of Ilocos Sur.

[47] See p. 7, note 1.

[48] Barrows, Census of the Philippine Islands, Vol. I, pp. 456 ff., 1903.

[49] Paul P. de La Gironiere, who visited the Tinguian in the early part of the nineteenth century, describes these ornaments as follows: "Their heads were ornamented with pearls, coral beads, and pieces of gold twisted among their hair; the upper parts of the hands were painted blue; wrists

adorned with interwoven bracelets, spangled with glass beads; these bracelets reached the elbow and formed a kind of half-plaited sleeve. La Gironiere, Twenty Years in the Philippines, pp. 108 ff.

[50] See Cole and Laufer, Chinese Pottery in the Philippines (Pub. Field Museum of Natural History, Vol. XII, No. 1).

[51] This is entirely in agreement with Chinese records. The Islands always appeared to the Chinese as an Eldorado desirable for its gold and pearls.

[52] See p. 17, note 2.

[53] See p. 7, note 4.

[54] A bamboo pole, about ten feet long, one end of which is slit into several strips; these are forced apart and are interwoven with other strips, thus forming a sort of basket.

[55] See Cole, Distribution of the Non-Christian Tribes of Northwestern Luzon (American Anthropologist, Vol. II, No. 3, 1909, pp. 340, 341).

[56] See p. 9.

[57] See p. 10, note 3.

[58] Among the Ifugao, the lowest of the four layers or strata which overhang the earth is known as Kabuniyan. See Beyer, Philippine Journal of Science, Vol. VIII, 1913, No. 2, p. 98.

[59] See p. 8.

[60] An Ifugao myth gives sanction to the marriage of brother and sister under certain circumstances, although it is prohibited in every day life. Beyer, Philippine Journal of Science, Vol. VIII, 1913, No. 2, pp. 100 ff.

[61] As opposed to the spirit mate of Aponitolau.

[62] According to Ling Roth, the Malanaus of Borneo bury small boats near the graves of the deceased, for the use of the departed spirits. It was formerly the custom to put jars, weapons, clothes, food, and in some cases a female slave aboard a raft, and send it out to sea on the ebb tide "in order that the deceased might meet with these necessaries in his upward flight." Natives of Sarawak and British North Borneo, Vol. I, p. 145, (London, 1896). For notes on the funeral boat of the Kayan, see Hose and McDougall, Pagan Tribes of Borneo, Vol. II, p. 35.—Among the Kulaman of southern Mindanao an important man is sometimes placed in a coffin resembling a small boat, which is then fastened on high poles near to the beach. Cole, Wild Tribes of Davao District, Mindanao (Pub. Field Museum of Natural History, Vol. XII, No. 2, 1913).—The supreme being, Lumawig, of the Bontoc Igorot is said to have placed his living wife and children in a

log coffin; at one end he tied a dog, at the other a cock, and set them adrift on the river. See Jenks, The Bontoc Igorot, p. 203, (Manila, 1905); Seidenadel, The Language of the Bontoc Igorot, p. 502 ff., (Chicago, 1909).

[63] For similar omens observed by the Ifugao of Northern Luzon, see Beyer, Origin Myths of the Mountain peoples of the Philippines (Philippine Journal of Science, Vol. VIII, 1913, No. 2, p. 103).

[64] Page 3, note 2.

[65] See tale 22.

[66] For a discussion of this class of myths, see Waterman, Jour. Am. Folklore, Vol. XXVII, 1914, p. 13 ff.; Lowie, ibid., Vol. XXI, p. 101 ff., 1908; P. W. Schmidt, Grundlinien einer Vergleichung der Religionen und Mythologien der austronesischen Völker, (Wien, 1910).

[67] See p. 10, note 3.

[68] The Pala-an is third in importance among Tinguian ceremonies.

[69] Tale 58.

[70] This is offered only as a possible explanation, for little is known of the beliefs of this group of Igorot.

[71] See p. 11, note 1.

[72] Tale 68.

[73] Hose and McDougall, The Pagan Tribes of Borneo, Vol. II, p. 148, (London, 1912).

[74] Bezemer, Volksdichtung aus Indonesien, p. 304, Haag, 1904. For the Tagalog version of this tale see Bayliss, (Jour. Am. Folk-lore, Vol. XXI, 1908, p. 46).

[75] Evans, Folk Stories of British North Borneo. (Journal Royal Anthropological Institute, Vol. XLIII, 1913, p. 475).

[76] Folk Stories of British North Borneo (Journal Royal Anthropological Institute, Vol. XLIII, p. 447, 1913).

[77] Tale No. 89.

[78] Hose and McDougall, The Pagan Tribes of Borneo, Vol. II, pp. 144-146.

[79] Tale 91. The cloak which causes invisibility is found in Grimm's tale of the raven. See Grimm's Fairy Tales, Columbus Series, p. 30. In a Pampanga

tale the possessor of a magic stone becomes invisible when squeezes it. See Bayliss, (Jour. Am. Folk-Lore, Vol. XXI, 1908, p. 48).

[80] Ratzel, History of Mankind, Vol. I, Book II. Graebner, Methode der Ethnologie, Heidelberg, 1911; Die melanesische Bogenkultur und ihre Verwandten (Anthropos, Vol. IV, pp. 726, 998, 1909).

[81] See Waterman, Journal American Folklore, Vol. XXVII, 1914, pp. 45-46.

[82] See Waterman, Journal American Folklore, Vol. XXVII, 1914, pp. 45-46.

[83] See Waterman, Journal American Folklore, Vol. XXVII, 1914, pp. 45-46.

[84] Stories of magic growth are frequently found in North America. See Kroeber, Gross Ventre Myths and Tales (Anthropological Papers of the Am. Mus. of Nat. Hist., Vol. I, p. 82); also Lowie, The Assiniboin (ibid., Vol. IV, Pt. 1, p. 136).

[85] Other examples of equally widespread tales are noted by Boas, Indianische Sagen, p. 852, (Berlin, 1895); L. Roth, Custom and Myth, pp. 87 ff., (New York, 1885); and others. A discussion of the spread of similar material will be found in Graebner, Methode der Ethnologie, p. 115; Ehrenreich, Mythen und Legenden der südamerikanischen Urvölker, pp. 77 ff.; Ehrenreich, Die allgemeine Mythologie und ihre ethnologischen Grundlagen, p. 270.

[86] Cole and Laufer, Chinese Pottery in the Philippines (Publication Field Museum of Natural History, Anthropological Series, Vol. XII, No. 1, Chicago, 1913).

[87] Nieuwenhuis, Kunstperlen und ihre kulturelle Bedeutung (Int. Arch. für Ethnographie, Vol. XVI, 1903, pp. 136-154).

[88] Philippine Journal of Science, Vol. III, No. 4, 1908, pp. 197-211.

End of Project Gutenberg's A Study in Tinguian Folk-Lore, by Fay-Cooper Cole